Sources of Alchemy and Chemistry
Sir Robert Mond Studies in the History of Early Chemistry

AMBIX VOLUME 70 SUPPLEMENT 1 2023

The Book on Alums and Salts of Pseudo-Rāzī

The Arabic and Hebrew Traditions

Gabriele Ferrario

Ambix (Print ISSN 0002-6980, Online ISSN 1745-8234) is published by Taylor & Francis Group, LLC, 4 Park Square, Milton Park, Abingdon, Oxon, OX14 4RN, UK. Periodicals postage paid (Permit no. 000-438) at Philadelphia, PA and additional mailing offices.

Subscription information: For more information and subscription rates, please see tandfonline.com/pricing/journal/YAMB. Taylor & Francis journals are available in a range of different packages, designed to suit every library's needs and budget. This journal is available for institutional subscriptions with online only or print & online options. This journal may also be available as part of our libraries, subject collections, or archives. For more information on our sales packages, please visit: librarianresources.taylorandfrancis.com.

For support with any institutional subscription, please visit help.tandfonline.com or email our dedicated team at subscriptions@tandf.co.uk.

Subscriptions purchased at the personal rate are strictly for personal, non-commercial use only. The reselling of personal subscriptions is prohibited. Personal subscriptions must be purchased with a personal check, credit card, or BAC/wire transfer. Proof of personal status may be requested.

Society membership: Individuals who wish to subscribe can join the Society for the History of Alchemy and Chemistry to receive the journal as part of their annual membership. For more information please visit www.ambix.org or contact the Membership Secretary, Dr Carolyn Cobbold at newjoiner@ambix.org

DOI: 10.1080/00026980.2023.2222505
ISBN (pbk) 978-1-032-86940-7

This publication is part of the research project *Alchemy in the Making: From Ancient Babylonia via Graeco-Roman Egypt into the Byzantine, Syriac, and Arabic Traditions*, acronym *AlchemEast*. The *AlchemEast* project has received funding from the European Research Council (ERC) under the European Union's Horizon 2020 research and innovation programme [grant agreement no. 724914].

European Research Council

AlchemEast

This volume is the second issue of Sources of Alchemy and Chemistry: Sir Robert Mond Studies in the History of Early Chemistry: a series of publications that presents critical editions of core texts in the history of alchemy and early chemistry, together with English translations, commentary, and scholarly apparatus. The series is produced and distributed under the aegis of *Ambix* and the Society for the History of Alchemy and Chemistry, and named in honour of the Society's first and only President, Sir Robert Mond (1867–1938).

Cover image: MS Orient. Oct. (klein) 514, Staatsbibliothek zu Berlin – Preussischer Kulturbesitz, Orientabteilung, fol. 73 verso (detail).

Society for the History of Alchemy and Chemistry

Routledge
Taylor & Francis Group

Published by Routledge for the Society for the History of Alchemy and Chemistry

To Rima and Taos

CONTENTS

Acknowledgements

If the trite analogy between childbirth and the publication of a book could still be invoked to some effect, I am sure this monograph would be a long overdue baby, one that comes to life already in its teens. My research on the *On Alums and Salts* goes back to my student years at the University of Venice and was the object of the PhD dissertation that I discussed an embarrassing number of years ago at that same university. The present work is in many ways an updated, revised and expanded English version of my doctoral dissertation. It has taken on the current shape while travelling with me and my family between nine different homes, three countries and two continents. In all these years and places, I have been mentored, helped, supported, and encouraged by the most excellent people, whose trust in this work never wavered (or, at least, they never told me it did). Without them this book would have never been completed.

Robert Temple has generously provided support for the production of this monograph: his passion for historical research in the relationship between humans and nature sees no cultural, chronological or geographical boundaries; on more than one occasion, I have appreciated his historical curiosity and wide expertise, together with his and Olivia's exquisite hosting charm.

My deepest gratitude goes to Jennifer Rampling and Lawrence Principe, who are the mentors and editors every academic should find along their path; their immense knowledge, generosity, enthusiasm, and patience are the lymph that runs throughout these pages that they have read and revised numerous times, providing me with invaluable comments and suggestions. I am deeply indebted to Sébastien Moureau, who read and reviewed the edition and translation of the Arabic text, helped me solve some of its difficult passages, and saved me from several mistakes: his rare philological zeal is only matched by his academic generosity and kindness. I have treasured the comments and corrections of the anonymous reviewers who read a draft of this book: I am deeply thankful for their expertise and thorough effort that improved my work significantly.

Piero Capelli, Marina Rustow, Paolo Lucca, and Angelo Scarabel have read earlier versions of my work, provided very valuable feedback and kindly prodded me to seek its publication. Charles Burnett has been my mentor during a short-term fellowship at the Warburg Institute in London: our conversations on the Arabic and Latin *On Alums and Salts* in his office and in the canteen of the Institute still resonate in my ears and are mirrored in more than one passage of this book. With Anke Timmermann and Ron Brashear I conversed fruitfully about my work during a fellowship at the Chemical Heritage Foundation (now Science History Institute) in Philadelphia. I am very thankful to Ben Outhwaite, Esther-Miriam Wagner, Melonie Schmierer-Lee, Amir Ashur, Daniel Davies, Ronny Vollandt, Julia Krivoruchko, Zvi Stampfer, and all the researchers that passed through the Genizah Research Unit in Cambridge, UK, while I worked there: what I have learned from them in the Unit day by day had a great impact on this monograph.

The Department of History of Science and Technology at Johns Hopkins University in Baltimore has been for me a welcoming and stimulating home for three years: the weekly seminars of the department and my conversations with Maria Portuondo, Yulia Frumer, Joris Mercelis, Nuno Castel-Branco, and the excellent group of graduate students have opened my eyes to implications of my research and of the topic of the present book for the history of science more at large. Of course, my Baltimorean experience would have not been possible without the mentorship and help of Lawrence Principe, who deserves many more praises than the two mentions that I managed to fit in this space. In the ERC AlchemEast Project at the University of Bologna, I found the perfect environment for completing this monograph and a group of likeminded colleagues, whose work on alchemy across times and cultures keeps on having a fundamental influence on my ideas about alchemy. Matteo Martelli, the best PI, colleague, and friend on earth, Lucia Raggetti, Vincenzo Carlotta, and the other members of the team have been erudite and friendly companions during the final stages of the preparation of this work: I have treasured their useful feedback and punctual comments on portions of this book, whose final form owes much to their insights. The annual workshops of the so-called "Sources Group" are the most coveted academic events in my diary and have been the ferment that propelled the transmutation of my doctoral dissertation into a book: Sébastien Moureau, Bink Hallum, Marion Dapsens, Salam Rassi, and Sebastian Richter, together with the other members of the group mentioned above, have indefatigably sat through long sessions of work on the *On Alums and Salts* and helped me immensely; on more than one occasion, I would read my translation, while the others checked it against the Arabic, Hebrew, and Latin versions: I felt as if I were in a modern *Bayt al-Ḥikma*.

This book is dedicated to Rima and Taos, the gazelle and the peacock that fill with beauty every day of my life.

Transliteration systems

Arabic

ا	ā	ظ	ẓ
ء	’	ع	‘
ب	b	غ	ġ
ت	t	ف	f
ث	ṯ	ق	q
ج	ǧ	ك	k
ح	ḥ	ل	l
خ	ḫ	م	m
د	d	ن	n
ذ	ḏ	ه	h
ر	r	ة	h, t
ز	z	و	w, ū
س	s	ي	y, ī
ش	š	ى	ā, ‘a, ‘i
ص	ṣ	لا	lā
ض	ḍ	ال	al-
ط	ṭ		

Hebrew

א	ʼ
ב	b
ג	g
ד	d
ה	h
ו	w
ז	z
ח	ḥ
ט	ṭ
י	y
כ ך	k
ל	l
מ ם	m
נ ן	n
ס	s
ע	ʻ
פ ף	*p*
צ ץ	ṣ
ק	q
ר	r
שׂ	ś
שׁ	š
ת	t

List of illustrations

INTRODUCTION

1. *On Alums and Salts*, a "classic of chemistry"

1.1. Preamble

Medieval knowledge did not stay still: it moved incessantly between languages and cultures, often transmuting in the process. The knowledge of alchemy was no exception, although its paths have been less well trodden by modern scholars than, say, mathematics or medicine. Yet, among the medieval arts and sciences, alchemy offers a particularly interesting example of knowledge in transit. Its practitioners sought to preserve the knowledge of their own authoritative forebears, not all of whom actually wrote the works later attributed to them, while at the same time seeking to interpret those works in light of their own experience. For practitioners of alchemy, ancient works might appear extremely relevant to their own practical concerns, even as they worked to reinterpret and reshape their contents.[1]

The twelfth century marked a watershed in the transmission of alchemical knowledge – prior to this, alchemy was essentially an unknown discipline in the Latin world. While some technical texts were already circulating in Europe, these works were not based on any explicit theoretical framework, nor did they demonstrate the theoretical maturity that alchemy had already achieved in the Graeco-Egyptian and Byzantine world and the Islamic lands.[2] During the twelfth century, however, this situation changed radically, thanks to the zealous activity of a group of translators working in the multicultural and multilingual context of Andalusia (al-Andalus) in southern Spain. This movement saw numerous philosophical and scientific works translated from Arabic into Latin, including writings on alchemy. The arrival of these materials prompted a host of new readers to investigate the exciting, novel discipline, and to study, copy, and comment upon the translated alchemical texts.[3] Latin translations were in turn rendered into vernacular languages, acquiring new commentaries and interpretations as they moved between communities and cultures. Such interventions ensured that neither alchemical texts, nor the knowledge they contained, remained static over time.

One of the most important and influential of these textual travellers was *The Book on Alums and Salts*. This famous treatise was often (incorrectly) attributed to the great tenth-century Persian physician and alchemist Muḥammad ibn Zakarīyā' al-Rāzī, a factor that undoubtedly contributed to its wide circulation.

[1] For an engaging overview of the history of alchemy from antiquity to modern times, see Lawrence M. Principe, *The Secrets of Alchemy* (Chicago: University of Chicago Press, 2013). For a general survey on the sources and main developments of alchemy in the Arabo-Islamic world, see George Anawati "L'alchimie arabe," in *Histoire des sciences arabes*, ed. Roshdi Rashed, 3 vols. (Paris: Seuil 1997), III, 111–41.

[2] These works include the *Mappae Clavicula*, *Compositiones Lucenses*, and *Liber Ignium*.

[3] The period of cultural efflorescence determined by the translation of Arabic philosophical and scientific works has been described as the "twelfth-century Renaissance": Charles H. Haskins, *The Renaissance of the Twelfth Century* (Cambridge: Harvard University Press, 1927). On the survival and assimilation Greek scientific culture in medieval Islamic thought, see Dimitri Gutas, *Greek Thought, Arabic Culture: The Graeco-Arabic Translation Movement in Baghdad and Early 'Abbāsid Society* (Oxford: Routledge, 1998).

It was also highly regarded among medieval and early modern alchemical practitioners for its accomplished synthesis of alchemical theory and laboratory practice. These qualities allowed *On Alums and Salts* to successfully traverse linguistic, geographical, and religious borders. Originally written in Arabic, the work achieved its greatest popularity in Christian Europe, thanks to the success of its Latin translation, the *Liber de aluminibus et salibus*. Later, an anonymous Jewish alchemist would draw upon the Arabic and possibly also the Latin versions to produce a new, expanded Hebrew translation that seems to blend both traditions, while also incorporating the fruits of his own practical experience and cultural milieu. As a result, *On Alums and Salts* – henceforth *OAS* – offers an exemplary case study for mapping how both knowledge and text transform.[4]

The combination of authority and practical know-how that appealed to the work's medieval readers continues to spur interest in the *OAS* today. The *OAS* is first and foremost a practical handbook of alchemical recipes, whose author chose to avoid the obscurity and obfuscation that characterise much medieval and early modern alchemical literature. Instead, the aims of the recipes that constitute the bulk of material in *OAS* are clearly stated. Ingredients and their quantities are listed in an orderly fashion, while cover-names – the so-called *Decknamen* under which alchemists traditionally disguised the substances used in their works – are rarely used and relatively easy to decipher.[5] Laboratory operations and apparatus are described in a clear and straightforward style, without the imaginative Pindaric flights that students might expect to find in alchemical authorities. Modern readers are often struck by the work's clear organisational structure and practical content: features of that led one editor, Robert Steele, to define it as "a classic of chemistry."[6]

The aim of this volume is to foreground both the multilingual journey of the text and the practical significance of its chemistry, by presenting new editions of both the Arabic and the Hebrew versions of the *OAS*, together with facing-page English translations. In the past, scholarly attention has tended to focus primarily on the work's extensive Latin reception, an orientation characterised by the tendency in this scholarship – including my own earlier contributions to the topic – to refer to the text as *LAS*, the acronym of its Latin title. In order to do justice to all the linguistic and cultural manifestations of the work, in this volume I shall refer to it as *On Alums and Salts* (*OAS*), to reflect the title of the treatise as it appears in all three linguistic traditions.

This linguistic journey begins in Arabic – but in order to reposition the Arabic exemplar at the centre of the study of the *OAS*, a new edition is required that

[4] My preliminary studies on the *On Alums and Salts* have appeared in two articles: Gabriele Ferrario, "Il *Libro degli allumi e dei sali*: *status quaestionis* e prospettive di studio," *Henoch* 26 (2004): 275–96 and Gabriele Ferrario, "Origins and Transmission of the *Liber de aluminibus et salibus*," in *Chymists and Chymistry: Studies in the History of Alchemy and Early Modern Chymistry*, ed. L. Principe (Sagamore Beach, MA: Science History Publications, 2007), 137–48.

[5] On *Decknamen* in medieval Arabic alchemy, see Alfred Siggel, *Decknamen in der arabischen alchemistischen Literatur* (Berlin: Akademie-Verlag, 1951).

[6] Robert Steele, "Practical Chemistry in the 12th Century. Rasis de aluminibus et salibus," *Isis* 12 (1929): 10–49.

accurately represents the available material evidence. While the eminent philologist Julius Ruska previously edited the Arabic text in 1935, he was unable to consult the only surviving Arabic manuscript, Berlin Staatsbibliothek MS Sprenger 1908, and he also chose not to reproduce the order of material as it appears in this manuscript.[7] In an effort to present a text that is as close as possible to the manuscript tradition, I have preserved the order of sections as found in the Arabic witness, as well as providing the first English translation of the text.

The Hebrew tradition, preserved in the unique manuscript Berlin Staatsbibliothek MS Orient. Oct. klein 514, has not previously been edited or translated. As we shall see, the path of this translation was far from a linear journey. While the translator worked primarily from an Arabic manuscript, it is possible that he also consulted a Latin version of the text, while making extensive use of Italian terms and phrases, and finally embellishing the whole with his own practical observations. The resulting work, both translation and commentary, offers new evidence for the reception of Arabic alchemy in early modern Europe, while at the same time offering a window into the alchemical practice of an otherwise unknown Jewish alchemist, whose notes reveal both respectful engagement with his textual authority, and careful testing and critique of its practical content. By presenting both editions in a single volume, together with the first published English translations of the *OAS*, I hope to provide readers with an opportunity not only to study the text of an important work of medieval alchemy, but also to trace how it was read and used – and how it changed.

In the pages that follow, I present the current state of the art regarding the *OAS*; the problems related to its attribution and title; the structure and alchemical content of the treatise (and its consequent appeal to medieval readers); the features of the Arabic and Hebrew manuscripts that preserve the text; and, as an appendix, an overview of the manuscript witnesses in the Latin tradition of the *OAS*. Other appendices include the tables of contents of the Arabic and Hebrew manuscripts and a multilingual lexicon of alchemical terminology that functions also as an index of the alchemical terms found in the edited texts.

1.2. Modern studies

The *OAS* has long attracted the attention of scholars, some of whom have been intrigued by the association with al-Rāzī, others by the operative nature of the treatise. As noted above, the various Latin versions of the *OAS* have received the lion's share of scholarly interest. However, the Arabic text became an object of serious study during the 1930s, while the 1990s shone a brief spotlight on the Hebrew translation. Before embarking on a new study of the rich, trilingual life of the *OAS*, it is therefore helpful to clear the ground by recalling this literature – in particular, the ground-breaking work of Julius Ruska, the first editor of the Arabic text.

[7] Paul Kraus, "Julius Ruska. Festschrift zum 70. Geburstag," *Osiris* 5 (1938): 4–40.

The first reference to a Latin version of the *OAS* appeared in Isidoro Carini's *Sulle science occulte nel Medio Evo* (On the Occult Sciences in the Middle Ages) published by the Sicilian printing house Perino in 1872.[8] In this book, Carini sought to trace a brief but very erudite history of the "occult sciences," particularly alchemy, by explaining and contextualising the alchemical content of a Latin manuscript miscellany, the so-called "Codex Speciale," which included an early witness of the *OAS*.[9] Carini noted, but did not question, this manuscript's peculiar pseudo-attribution, describing the *OAS* as a treatise attributed to Hippocrates and to Galen, with the title *Discourse on Alums and Salts That Are Needed in This Art*, as well as a further note, "Some title it so: 'Here begins the book of Hippocrates and Galen.'"[10]

The *OAS* received further attention from the chemist Marcelin Berthelot in the first volume of his work devoted to medieval alchemy. When describing the contents of the Paris manuscript Bibliothèque Nationale MS lat. 6514, Berthelot noticed a reference to al-Rāzī: "Incipit liber Rasis de aluminibus et salibus" ("Here begins al-Rāzī's book on alums and salts," fol. 128r). Berthelot characterised the treatise as "an essentially practical writing, where one finds some recipes that often deal with the same topics as those of previous works."[11] He also noted parallel passages between the Latin *OAS* and the discussion of minerals in the encyclopaedic works of Vincent of Beauvais (d. ca. 1264) – recognising that the presence of numerous variants meant that Vincent had probably worked from a different Latin version to that found in MS lat. 6514.[12]

The first edition of the Latin *OAS* appeared in 1929, when Robert Steele edited the same Parisian manuscript described by Berthelot.[13] He also collated it with another Latin version of the work (British Library, MS Arundel 164, fols. 167v-75v) and with the indirect tradition represented by quotations from the *OAS* preserved in medieval encyclopaedias. In addition to identifying numerous parallels between Vincent's citations and the text preserved in MS lat. 6514, Steele discovered that the *OAS* was also quoted as a source of mineralogical

[8] Isidoro Carini, *Sulle scienze occulte nel Medio Evo e sopra un codice della famiglia Speciale* (Sala Bolognese: A. Forni, 1983; reprint of Palermo: Perino, 1872).

[9] Carini's other goal in writing the monograph was to provide information on the contents of the "Codex Speciale" to bring the manuscript to the attention of the Municipal Library of Palermo, which Carini hoped would purchase it. He dedicated the book to Salvatore Cusa, then professor of palaeography with responsibility for teaching the Arabic language in Palermo. Carini's wish became reality and the "Speciale codex" is now held at the Biblioteca Comunale di Palermo (MS 4 Qq A 10). For a description of the codex and its history, see below, Appendix 4.

[10] Carini, *Sulle scienze occulte*, 27: "un trattato, che si aggiudica ad Ippocrate ed a Galeno *Sermo de aluminibus et salibus que in hac arte necessaria sunt* colla nota *Aliqui intitulant hunc ita. Incipit liber ypocratis et galieni*." The second part of Carini's monograph, the *Elenco dei trattati e dei capitoli contenuti nel Codice Speciale*, is a straightforward index of the content of the Speciale manuscript. Two pages of this section are devoted to a list of the Latin headings of the chapters of the *OAS* as preserved in the Speciale manuscript: Carini, *Sulle scienze occulte*, XIV–XV.

[11] Marcellin P. E. Berthelot, *La chimie au Moyen Âge*, 3 vols. (Osnabrück-Amsterdam: Zeller-Philo Press, 1967; reprint of first ed. Paris, 1893), I, 317–19: "C'est un écrit essentiellement pratique, et où se trouvent des recettes traitant fréquemment les mêmes sujets que celles des opuscules précédents."

[12] For the use of the *OAS* in medieval encyclopaedias, see below, Appendix 4.

[13] Steele, "Practical Chemistry."

knowledge in the *Opus Minus* of Roger Bacon (d. 1292), who – Steele argues – "used a much enlarged text."[14]

The most important study on the *OAS* to date was published by Julius Ruska in 1935.[15] In his monograph, Ruska offered an edition and German translation of a different and, according to him, more reliable Latin version of the work which he had discovered within the text of John of Garland's *Compendium alchemiae* (Basel, 1560).[16] Ruska's work also includes the first edition of the only extant witness of the Arabic original of the *OAS*, as preserved in MS Sprenger 1908 of the Berlin Staatsbibliothek.[17] In many ways, Ruska's introduction to his editions and translation still represents the state of the art regarding questions of the attribution, context, and character of the *OAS*, and will serve as a constant point of reference here.[18]

More recently, Raphael Patai devoted a chapter of his extensive survey of the Jewish contributions to alchemy to a Hebrew version of the *OAS* found in MS Orient. Oct. 514, also in the Berlin Staatsbibliothek.[19] Patai provided a brief codicological and linguistic description of the manuscript, an overview of the alchemical authorities there mentioned, and an English translation of selected passages.

A complete English translation of the Latin text remains a desideratum, although Italian translations of a generous selection of passages, based on Steele's edition, can be found in the elegant anthology of Western alchemical literature edited by Michela Pereira and Pinella Travaglia.[20] The complex task of disentangling the Latin tradition of the *OAS* and producing a critical edition was undertaken by the late Catherine Arbuthnott in her doctoral dissertation. Arbuthnott's work, which still awaits publication, represents a major contribution for our understanding of the relationship between the various Latin witnesses of the *OAS*, finally providing scholars with a critically established basis for studying the Latin *OAS* and its diffusion in Christian Europe.[21]

[14] Steele, "Practical Chemistry," 12. Steele enriched his article with a "Glossarial Index" that marks the occurrences of alchemical technical terms in the treatise and offers an English translation and brief explanation of this terminology. He notes the dependence of his glossary on Lucien Leclerc, *Traité des Simples par Ibn el-Beïthar. Notices et extraits des manuscrits de la Bibliothèque Nationale*, vols. XXIII, XXV, XXVI (Paris: Imprimerie Nationale, 1883).

[15] Julius Ruska, *Das Buch der Alaune und Salze: ein Grundwerk der Spätlateinischer Alchemie* (Berlin: Verlag Chemie, 1935).

[16] John of Garland (Lat. Johannes Garlandius, d. 1252), an Englishman educated at Oxford, moved after 1202 to France and worked at the University of Toulouse. He authored the *De Triumphis Ecclesiae*, an extensive poem where the victories of the church over heretics at home and infidels abroad are praised, edited in Thomas Wright, ed., *De Triumphis Ecclesiae Johannis de Garlandia* (London: Nichols, 1856). A number of treatises on grammar, maths and music are also attributed to John of Garland.

[17] For a detailed description of MS Sprenger 1908, see below, § 4.

[18] Four years later, Ruska returned to the debated problem of the paternity of the *OAS* in an article devoted to the pseudepigraphic writings attributed to al-Rāzī: Julius Ruska, "Pseudoepigraphe Rasis-Schriften," *Osiris* 7 (1939); on the *OAS* in particular, see 39–40.

[19] Robert Patai, *The Jewish Alchemists. A History and Source Book* (Princeton: Princeton University Press 1994), 119–24 (on the Hebrew *OAS*); 407–16 (on the miscellaneous manuscript that contains the *OAS*). For a detailed description of MS Orient. Oct. 514, see below, § 5.

[20] Michela Pereira, *Alchimia. I testi della tradizione occidentale* (Milan: Mondadori 2006), 272–305. Pinella Travaglia edited the Arabic contributions to this anthology.

[21] Catherine J. Arbuthnott, "Pseudo-Razi *De aluminibus et salibus*: A Critical Edition and Translation of the Latin Translation with Notes on the Chemical Procedures" (unpublished PhD thesis, University of London, 2002).

2. *On Alums and Salts and al-Rāzī, a problematic attribution*

When it first appeared in Latin Europe, *On Alums and Salts* was attributed to the tenth-century Persian physician and alchemist Muḥammad ibn Zakarīyā' al-Rāzī, and therefore was believed to have been composed during the tenth century in the region between Persia and Baghdad. This illustrious attribution, while it helped establish a wide circulation for the treatise in the Latin world, is nevertheless problematic. Al-Rāzī is not mentioned as author in either the Arabic or Hebrew manuscripts of the *OAS*, or in the vast majority of its Latin witnesses. In fact, the attribution only appears in a single surviving copy, Bibliothéque Nationale MS Lat. 6514, which names the author as "Bubacher Magument fil. Ceceri Arrasi."[22] How, then, did the attribution to al-Rāzī become so widely accepted in medieval Europe?[23]

The association between the great physician and the *OAS* appears to date to the twelfth century, when Gerard of Cremona (d. 1187) translated the Arabic treatise into Latin. We cannot determine with certainty whether Gerard found the name of al-Rāzī in the Arabic manuscript he translated, or if he was responsible for this attribution himself. However, the *OAS* appears (with the attribution to al-Rāzī) among the alchemical works listed in the *Commemoratio librorum*, a catalogue of Gerard's translations compiled by members of his circle, which circulated together with Gerard's *Vita* and *Eulogium*. Other alchemical translations by Gerard are mentioned in the catalogue, including the Jabirian *Liber divinitatis* and another pseudo-Rāzian treatise titled *Liber luminis luminum*.[24] Among surviving evidence, the Paris manuscript MS Lat. 6514 is believed to preserve a witness of Gerard's translation, representing the very beginning of the Latin reception of the work.

The fact that the *OAS* circulated under al-Rāzī's name provided surely a boost to the treatise's fortune, since al-Rāzī was a well-respected and avidly read author of medical and alchemical works which enjoyed wide diffusion and significant prestige in the Latin Middle Ages. This prestige undoubtedly enhanced the authority and dissemination of those texts spuriously attributed to him, especially in the case of the *OAS*, given its apparent translation by such a respected figure as Gerard of Cremona. Indeed, if we were to base our judgement solely on the extant manuscript witnesses of the *OAS* – disregarding all the random factors that determined the survival of those specific manuscripts, Arabic as well as Latin – we might conclude that the work was far more successful in its Latin translation than in its original language.

22 On this manuscript and the other Latin witnesses of the *OAS*, see below, Appendix 4.

23 On the original alchemical works by al-Rāzī, see Julius Ruska, "Al-Razi (Rhazes) als Chemiker," *Zeitschrift für Angewandte Chemie* 35 (1922): 719–24; Julius Ruska, "Al-Razi als Bahnbrecher einer neuer Chemie," *Deutsche Literatur-Zeitung* 44 (1923): 117–24; Rudolf Winderlich, "Ruska's Researches on the Alchemy of Al-Razi," *Journal of Chemical Education* 13 (1936): 313–15; Gerard Heym, "Al-Razi and Alchemy," *Ambix* 1 (1938): 184–91; James R. Partington, "The Chemistry of Al-Razi," *Ambix* 1 (1938): 192–96.

24 The critical edition of Gerard's *Vita*, *Eulogium*, and *Commemoratio* and their English translations was published in Charles Burnett, "The Coherence of the Arabic-Latin Translation Program in Toledo in the 12th Century," *Science in Context* 14 (2001): 249–88. The reference to the *OAS* among Gerard's translations is on 280 (n. 65).

This success was cemented in the thirteenth century, when, as noted above, Vincent of Beauvais and Roger Bacon adopted the *OAS* as one of their main sources of chemical information. Their decision to quote from the *OAS* in their own influential writings testified to the importance of the treatise, enhancing its aura of authority and contributing to its further diffusion.

Vincent of Beauvais seconded Gerard's opinion on the paternity of the text, using the chemical portions of the *OAS* extensively in his encyclopaedic *Speculum maius*, one of the most important medieval compilations of natural knowledge. The *OAS* is Vincent's major source on alchemical substances throughout the various volumes of the *Speculum*, as set down in Table 1 below. While the *bifaria* version of Vincent's encyclopaedia (completed before the year 1244) preserves only one quotation from the *OAS* (on the topic of tin, *Bif.* 5, 91c), the *trifaria* version (completed around the year 1259) quotes the *OAS* 32 times. These references are particularly dense in the seventh book of the *Speculum naturale*, which is devoted to the third day of Creation and deals, therefore, with mineralogy, and in the eleventh book of the *Speculum doctrinale*, which is devoted to the mechanical arts.[25] Vincent repeatedly names his source, introducing citations with such phrases as "Out of the book of alums and salts" (*Ex libro de aluminibus et salibus*), "Razi in the book of alums and salts" (*Razi in libro de aluminibus et salibus*), "Razi in the book of alums" (*Razi in libro de aluminibus*), and "Out of Razi's book on alums and salts" (*Ex libro Razi de aluminibus et salibus*).

Interestingly, Vincent selected his quotations primarily from the more theoretical passages of the *OAS*, namely those portions of text that introduce the sections on particular alchemical ingredients, and which describe their nature and uses while also providing general directions for working with them. The encyclopaedist therefore seems to have privileged general and theoretical material over the recipes and practical directions that make up the majority of the *OAS*, but which Vincent himself did not use. As Moureau also notes, Vincent's quotations are not always literal or complete: rather, he extracts portions from the *OAS*'s account of particular ingredients that serve his needs in the *Speculum*, while skipping sections that preserve non-alchemical material.[26] The table below records all the passages from the *OAS* quoted by Vincent in the *Speculum naturale* and *Speculum doctrinale*, illustrating the range of chemical substances covered in these works.[27]

The scholar and Franciscan friar Roger Bacon also made use of the *OAS* in the alchemical sections of his *Opus minus*, although without explicitly attributing it to al-Rāzī. As noted by Roger's nineteenth-century editor, Steele, the quotations

[25] The most complete and up-to-date study of the alchemical sources used by Vincent of Beauvais is Sébastien Moureau, "Les sources alchimiques de Vincent de Beauvais," *Spicae. Cahiers de l'Atelier Vincent de Beauvais* (New Series) 2 (2012): 5–118; the discussion on the *OAS* is found on 21–24, and Vincent's quotations on 65–72. My notes here are based on Moureau's exhaustive work. Ruska, *Das Buch*, 14, and Steele, "Practical Chemistry," 12, also provide tables of the correspondences between sections of the *OAS* and the chapters in Vincent's *Specula* in which they are quoted.

[26] Moureau, "Les sources," 23.

[27] I am here referring to the foliation of the Douai edition of the *Speculum*.

TABLE 1.

USE OF THE *OAS* BY VINCENT OF BEAUVAIS.

Topics in *OAS*	Cited in *Speculum naturale*	Cited in *Speculum doctrinale*
On vitriol	*Spec. Nat.* VII 75a	*Spec. Doct.* XI 122a1
On alum	*Spec. Nat.* V 94d	*Spec. Doct.* XI 122a2
On salt	*Spec. Nat.* V 86a	*Spec. Doct.* XI 118a1
On glass	*Spec. Nat.* VI 79a	*Spec. Doct.* XI 121a
On arsenic	*Spec. Nat.* VII 70a	*Spec. Doct.* XI 118a2
On sal ammoniac	*Spec. Nat.* VII 72c	*Spec Doc* XI 118a1
On sulphur	*Spec. Nat.* VII 67a	*Spec. Doct.* XI 118a3
On mercury	*Spec. Nat.* VII 62a	*Spec. Doct.* XI 119b
On the composition of metals	*Spec. Nat.* VII 6b	*Spec. Doct.* XI 105c
Congelation of metals	*Spec. Nat.* VII, 90a	*Spec. Doct.* XI 127
On gold	*Spec. Nat.* VII 13b	*Spec. Doct.* XI 111a
On the preparation of metals	*Spec. Nat.* VII 91b	*Spec. Hist.* XI, 128b
On silver	*Spec. Nat.* VII 18b	*Spec. Doct.* XI 112a
On tin	*Spec. Nat.* VII 38a	*Spec. Doct.* XI 115a; *Bif.* V 91c
On lead	*Spec. Nat.* VII 42c	*Spec. Doct.* XI 116a
On iron	*Spec. Nat.* VII 54a	*Spec. Doct.* XI 114a
On copper	*Spec. Nat.* VII 26a	*Spec. Doct.* XI 113a

in his work present many variants from the known Latin witnesses, suggesting that he may have extracted them from a different version of the *OAS* that has not survived.[28]

Centuries later, Gerard's translation and Vincent's quotations seem to have exerted considerable influence on historians of chemistry. Ferdinand Hoefer's *Histoire de la Chimie* mentions the three alchemical works attributed to al-Rāzī that he found in the aforementioned MS Lat. 6514: the *Liber Raxis qui dicitur Lumen Luminum Magnum* ("al-Rāzī's book called The Great Light of the Lights"), the *Liber Perfecti Magisterii Rhasei* ("The Book of the Perfect Magistery by al-Rāzī"), and the *Liber Rasis de Aluminibus et Salibus, quae in hac arte sunt necessaria* ("al-Rāzī's Book on Alums and Salts, which are needed in this art"). Hoefer concludes his description of these works by stating that "nothing indicates that the three writings by al-Rāzī are apocryphal. There is no solid proof that can be asserted against their authenticity."[29] The same attribution was considered valid by the German chemist and historian of chemistry Hermann Kopp in 1875.[30]

[28] Steele, "Practical Chemistry," 12.

[29] Ferdinand Hoefer, *Histoire de la chimie depuis les temps les plus reculées jusqu'a notre époque*, 2 vols (Paris: Hachette, 1842–43), I, 323–25: "Rien n'indique que les trois écrits de Rhasès soient apocryphes. Il n'y aurait aucune prevue solide à faire valoir contre leur authenticité."

[30] Hermann Kopp, *Ansichten über die Aufgabe der Chemie und über die Grundbestandtheile der Körper bei den bedeutenderen Chemikern* (Braunschweig: Vieweg, 1875), 54–55, n. 63.

More recently, the prolific editor of medieval philosophical texts, Robert Steele, introduced his own first edition and translation of a Latin witness of the *OAS* by reiterating his conviction that the traditional attribution of the work was correct: "it purports to be, and no doubt is in substance, the work of one of the most celebrated of Eastern physicians, Muhammad Abu Bakr Zakarya Al-Razi, many of whose medical writings were published in the infancy of printing."[31] Steele was aware of a number of references to Spanish geographical locations in the Latin manuscript he was editing, which might have looked out of place in a text supposedly originating in the Middle East, but he explained their presence by arguing that the Arabic text had been significantly modified by Moorish alchemists, who added their own interpolations and additions before the work was translated into Latin.

Doubts regarding the attribution of the *OAS* to al-Rāzī started to surface around the time that Steele's edition was published. In the list of original works by al-Rāzī published by George Ranking in 1914 there is no mention of the *OAS*.[32] Dorothea Singer was more explicit in her catalogue of alchemical manuscripts in Great Britain and Ireland, observing that "it will be noticed that the ascription of this work to Rhazes came late."[33]

The most important and substantial criticism of the traditional attribution of the *OAS* was formulated by Ruska in 1935.[34] Ruska argued that neither Gerard of Cremona nor Vincent of Beauvais was in a position to judge the authenticity of an Arabic work and that, similarly, such judgment was not possible for the nineteenth-century historians of chemistry. Any proper discussion of the attribution of the work to al-Rāzī would need to be based on direct knowledge of original works by the Persian physician and alchemist – something that was not available to any of them. Ruska's criticism touches also upon Steele's suggestion regarding Moorish interpolations to the *OAS*: Steele was not in a position to quantify the interpolations and additions he proposes since we have no evidence regarding the text of *OAS* before those additions and changes would have been made. In the present state of our knowledge, Ruska's criticism appears well-founded, as does his proposal to understand the *OAS* as the work of an anonymous Arab alchemist operating in Spain sometime during the eleventh century.

Below, I will briefly outline the main points of Ruska's argument and provide the references to the Hebrew version of the *OAS*, whose existence was unknown to Ruska.

31 Steele, "Practical Chemistry," 10. I here preserve Steele's non-scientific transliteration.

32 George S.A. Ranking, "The Life and Work of Rhazes," *Proceedings of the XVIIth International Congress of Medicine (1913), Section of the History of Medicine* (Oxford, 1914), 246–68; reprinted in Fuat Sezgin, *Muhammad ibn Zakariya al-Razi (d. 313/925): Texts and Studies* (Frankfurt am Main: Institute for the History of Arabic-Islamic Science at Johann Wolfgang Goethe University, 1996).

33 Dorothea W. Singer, *Catalogue of Latin and Vernacular Alchemical Manuscripts in Great Britain and Ireland*, 3 vols. (Brussels: Lamertin, 1928), I, 107–8.

34 Ruska, *Das Buch*, 15–18.

1. Only the scribe of BNF MS Lat. 6514 and Vincent of Beauvais (both of whom may rely on the same exemplar) attribute the work to al-Rāzī; all other witnesses are anonymous, including – I may add – the Hebrew translation.

2. In the *Kitāb al-Asrār*, an authentic alchemical treatise by al-Rāzī (which Ruska calls *Sirr al-Asrār*, causing some confusion) sal ammoniac is classified among the spirits, while in the *OAS* it is included and highly-praised among salts.[35]

3. The discussion of different kinds of salts includes numerous references to Spanish geographical locations which would very likely have been unknown to al-Rāzī.[36]

4. In the description of the best quality vitriol (unfortunately not preserved in the Arabic manuscript) further references to Spain are present: fol. 19 of the Hebrew text reads: "the best is with us in the island of Spain that they know as Sevilla"; § 1 of Steele's edition has: "quod asportatur de Ebla" ("which is extracted from Ebla"); and the corresponding § 72 of Ruska's edition of the Latin text has "quod adducitur de Labla insula" ("which is brought from the Labla island"). According to Asin Palacios, Ebla/Labla should be identified with the city of Niebla, located in the Huelva province some sixty kilometres west of Seville.[37] The thirteenth-century Persian geographer al-Qazwīnī reports that this city is close to Seville and that a river (possibly identifiable with today's Rio Tinto) flows in its proximity. This river, al-Qazwīnī adds, springs from three sources: one of sweet water, one of water rich in vitriol, and a third of water rich in alum.[38] The seventeenth-century biographer al-Maqqarī also noticed that a very famous source for the extraction of vitriol was found in the city of Niebla.[39]

5. In the discussion of different alums, in a passage that is unfortunately not present in the Arabic manuscript, the Hebrew *OAS* reads (fol. 20r): "Ben Yulio from Cordoba said that its mines are in the midday – this means south – in the land of Cordoba called Anamariym." The corresponding passage in Steele's edition (§3) reads "dixit filius Gilgil Cordubensis quod ei erat minera a septent[r]ione Cordube in loco qui dicitur Neerim, ex circuito Cordube veteris" ("the son of Gilgil of Cordoba said that its mines were from the north of Cordoba in a place called Neerim, in the surrounding of the old Cordoba") while in Ruska's edition (§74) we find "et dixit fil. Inthuelis Cordubensis quod alumen habet mineras contra meridiem, in territorio Cordubae in quodam loco, qui vocatur Agnamarasin" ("and the son of Inthuelis of Cordoba said

[35] Julius Ruska, "Übersetzung und Bearbeitung von al-Razi's *Buch Geheimnis der Geheimnisse*," in *Quellen und Studien zur Geschichte der Naturwissenschaften* IV (Berlin: Springer, 1935), 153–238. An English translation of this work was made based on Ruska's German translation: Gail Marlow Taylor, *The Alchemy of Al-Razi. A Translation of the Book of Secrets* (Self-published and printed, 2014).

[36] For the geographical names in the Arabic *OAS*, see below, § 4.3.

[37] Miguel Asin Palacios, "Notes to R. Steele's Edition of Rasis *de aluminibus et salibus*," *Isis* 13 (1929–30): 358–9.

[38] For al-Qazwīnī, see Tadeusz Lewicki, "Al-Ḳazwīnī," in *Encyclopédie de l'Islam 2* (Leiden: Brill, 1960–96), IV, 898–900, and Julius Ruska, "Kazwini-Studien," *Der Islam* 4 (1913): 14–66; 236–62. For the passage on Niebla, see al-Qazwīnī, *Géographie/Kosmographie: Aṯār al-Bilād*, ed. Ferdinand Wüstenfeld (Göttingen: Verlag der Dieterischen Buchhandlung, 1848), 372 (quoted by Asin Palacios).

[39] Al-Maqqarī, *Analectes sur l'histoire et la littérature des Arabes d'Espagne*, eds. Reinhart P. A. Dozy, Gustave Dugat, Ludolf Krehl, and William Wright (Leiden: Brill, 1855–61), I, 123 (quoted in Asin Palacios). On al-Maqqarī, see Évariste Lévi-Provençal and Charles Pellat, "Al-Maḳḳarī," in *Encyclopédie de l'Islam* 2, VI, 170–2.

that alums have mines towards the South, in the area of Cordoba, in a place called Agnamarasin"). Ben Yulio/Filius Gilgil/Filius Inthuelis can be identified as the Cordovan physician Abū Dāwūd Sulaymān b. Ḥasan al-Andalusī ibn Ǧulǧul (d. 994), who authored a work on the history of medicine and on famous physicians, called *Ṭabaqāt al-aṭibbā' wa-'l-ḥukamā'* (History of Physicians and Sages).[40] Even if it has not yet been possible to identify a locality in modern Spain that corresponds to this geographical indication, the passage is clearly referring to the area surrounding the city of Cordoba.

6. A particular laboratory tool, described as a kind of iron disk, also points to Spain. The Hebrew translation (fol. 33r) describes it as "an iron vase called *qqwl* in Toledo," while Ruska's Latin edition (§ 56) reads "in disco ferreo, quae [*sic*] Coleti dicitur leachola" ("In an iron disk, which in Coleto is called leachola").

These arguments put forward by Ruska to support the identification of Andalusia as the likely site of composition of *OAS* are very convincing, and currently there is no evidence in support of a different origin for the work.

Indeed, if we were to base our judgment on purely quantitative considerations, there is no reason to give the al-Rāzī attribution any more weight that we would grant to alternative proposals – after all, the physician's name is preserved only in one extant Latin manuscript. The only extant Arabic witness of the *OAS* – the version edited in this book – appears in a manuscript that explicitly claims to transmit an alchemical work by Mu'ayyid al-Dīn Abū Ismā'īl Al-Ṭuġrā'ī (d. 1121), the alchemist and *wazīr* of the Seljuks.[41] This example reminds us that, if we were to privilege the Arabic tradition over the fame that *OAS* acquired in Latin, we might well end up hailing the *OAS* as a work of pseudo-Ṭuġrā'ī, rather than pseudo-Rāzī.

3. *On Alums and Salts*, the tradition of an Arabic alchemical handbook

3.1. A fluid title

The problematic authorship of the *OAS* is not the only element of fluidity in the traditional history of this work. Although it is by the Latin title *De aluminibus et salibus* that the work became universally known, the work circulated under many different names – or even under no title – in medieval and early modern Europe.

Unfortunately, there is not enough evidence in the unique Arabic manuscript, MS Sprenger 1908, to determine the original title of the treatise. In it, the text of the *OAS* begins abruptly at fol. 19r signalled only by the phrase *al-qawl fī al-milḥ* ("The Discourse on Salt") with no graphical distinction from the continuous Arabic text that precedes and follows it. We should therefore consider this phrase to be a section title rather than the overarching title for the whole work: parallel section titles also open the chapter on salts in the Latin and Hebrew versions.

[40] Ibn Ǧulǧul, *Les generations du médicins et des sages*, ed. Fu'ād Sayyid (Frankfurt am Main: Institute for the History of Arabic-Islamic Science, 1996; reprint of the edition Al-Qāhira, 1955). On Ibn Ǧulǧul, see Albert Dietrich, "Ibn Djuldjul," in *Encyclopédie de l'Islam* 2, III, 778–79.

[41] On al-Ṭuġrā'ī and its relevance to the Arabic manuscript of the *OAS*, see below, 4.1.

I would therefore argue that the general title of the work – if one ever existed – did not survive into the Arabic version now available to us.

The manuscripts preserving Latin translations show a rather wide array of titles, as do mentions of the *OAS* in other medieval writings, including the references by Vincent of Beauvais, Roger Bacon, and John of Garland, summarised in Table 2 below.[42] For instance, Roger Bacon mentions the *OAS*'s canonical title only once in the *Opus maius*, abbreviating it to *De aluminibus* (p. 381). In other places he calls the work *De spiritibus et corporibus* ("Concerning Spirits and Bodies"), expressly stating that this is a valid alternative title for the treatise.[43]

TABLE 2.

ALTERNATIVE TITLES OF THE *OAS*.

Textual witness (in approx. chronological order)	Latin title of *OAS*	English translation
Palermo, Biblioteca Comunale, MS 4 Qq A 10 (main text) (marginal note)	*Sermo de aluminibus et salibus, quae in haec arte necessaria existunt* *Alii intitulant hunc ita Incipit Liber Ypocratis et Galieni*	The Discourse on Alums and Salts Needed in this Work Others title it thus: Here begins the Book of Hippocrates and Galen
Roger Bacon, *Opus minus*	*De spiritibus et corporibus*	On Spirits and Bodies
John Garland, *Compendium alchemiae*	*Tractatus de salium, aluminumque varietate, compositione et usu Scriptoris incerti*	Treatise on the Variety, the Composition and the Use of Salts and Alums by an Unknown Author
Paris, Bibliothèque Nationale, MS Lat. 6514	*Liber Rasis de aluminibus et salibus quae in hac arte sunt necessaria*	The Book by Al-Rāzī on Alums and Salts Needed in this Work
Oxford, Bodleian Library, MS Digby 119 (title in margin, possibly in hand of John Dee)	*Liber rationum super corpora, spiritus, sales et atramenta*	The Book of Doctrines Regarding Bodies, Spirits, Salts and Vitriols
Oxford, Corpus Christi College, MS 125	*Liber salium et alluminum alchemiae*	The Book of the Salts and Alums of Alchemy
Jena, Thüringer Universitäts- und Landesbibliothek, MS El. q. 20	*Liber lumen luminum*	The Book of the Light of Lights
Bologna, Biblioteca Universitaria, MS 474 (830)	*Sermo de aluminibus et salibus et atramentis et alii quae in hac arte necessaria existunt ad complementum eleizir*	The Discourse on the Alums, Salts, Vitriols, and Other Things Needed for Completing the Elixir
London, British Library, MS Arundel 164 (main text) Explicit	*Tractatus de aluminibus et salibus et primo de atramento* *Explicit secretum philosophiae secretissimum*	The Treatise on Alums and Salts and Firstly on Atrament "Here ends the most concealed secret of philosophy"

As the table shows, some manuscript witnesses include more than one title, while in others the title has been added by a later hand – in the case of MS Digby 119, possibly that of the English natural philosopher (and prolific collector of medieval alchemica) John Dee.[44] Despite the wide variation, these examples are generally

[42] For an overview of the Latin tradition of the *OAS*, see below, Appendix 4.

[43] These further quotations are found at pp. 376, 379, 380, 381, 382 in Steele's edition: Roger Bacon, *Opera hactenus inedita. Opus Minus, Opus Tertium, Compendium Philosophiae* (London: J. S. Brewer, 1859).

[44] Ruska, *Das Buch*, 16, suggests that John Dee might have been the owner of this manuscript. Dee (1527–1609) was a prominent English astronomer, astrologer, mathematician, and philosopher, who at various times received

consistent in referring to alums and salts. It is significant that the standard title by which the *OAS* is most widely known in Latin, the *Liber de aluminibus et salibus*, is echoed in the Hebrew version of the text: *Atḥil Sefer alumei we-melaḥim* ("I begin the Book on Alums and Salts").[45] There is therefore plenty of precedent for adopting *On Alums and Salts* as a generic title for the text across its many permutations.

3.2. Reconstructing the chemistry of *On Alums and Salts*

While Robert Steele identified the *OAS* as a classic work of chemistry, the categories of "alchemy" and "chemistry" were not clearly distinguished in the Middle Ages.[46] In this volume, I take the premodern approach to such terms as alchemy, chemistry, and chymistry, namely to view them all as referring to similar intellectual and practical endeavours, aimed at understanding the workings of nature, the behaviour of matter, and the ways in which materials could be manipulated. In recent decades, scholars have paid renewed attention to the materiality of premodern scientific knowledge – enquiries that find fertile ground within the pages of *OAS*. The author's straightforward descriptions of laboratory operations invite replication in modern laboratories, a methodology whose usefulness is becoming increasingly apparent to scholars of alchemy, and indeed of medieval technical knowledge more broadly, for whom careful, historically sensitive reconstruction offers a valuable exegetical tool.[47]

In both content and organisation, the *OAS* offers plenty of scope for such endeavours. The work is divided into textual units, each one opening with a theoretical passage that functions as an introduction to, and justification for, the various operations described in the work. For every chemical substance under discussion, the author provides a brief and incisive passage that describes its qualities, physical features, and general value to alchemists who intend to use that substance in their work. Together, these substances encompass virtually all the ingredients to be found in an alchemical workshop, including "bodies" (metals), "spirits" (volatile substances), and other materials. Although a number of the recipes in the *OAS* are ultimately concerned with gold-making (chrysopoeia) or silver-making (argyropoeia), not all of them are aimed at accomplishing the transmutational *opus*: very frequently, the procedures described in this treatise are preliminary steps used to

44 *Continued*
patronage from Queen Elisabeth I. Dee owned one of the largest private libraries of his time, which was partly catalogued by Dee himself: see Julian Roberts and Andrew G. Watson, *John Dee's Library Catalogue* (London: Bibliographical Society, 1990).

45 Fol. 19r. For a detailed description of the Hebrew manuscript and its contents, see below, § 5.

46 On the difficulty of using these terms, see William R. Newman and Lawrence M. Principe, "Alchemy vs. Chemistry: The Etymological Origins of a Historiographic Mistake," *Early Science and Medicine* 3 (1998): 32–65.

47 Lawrence Principe and Jennifer Rampling have already attempted to reconstruct several recipes from the *OAS*: see Jennifer M. Rampling, "How to Sublime Mercury: Reading Like a Philosopher in Medieval Europe," *The Recipes Project* (24 May 2018), https://recipes.hypotheses.org/10663. For an overview of reconstruction as a methodology, as well as several valuable case studies, see Hjalmar Fors, Lawrence M. Principe, and H. Otto Sibum, eds., "From the Library to the Laboratory and Back Again: Experiments as a Tool for Historians of Science," *Ambix* 63.2 (2017). For more recent reflections, see Sven Dupré, Anna Harris, et al., eds., *Reconstruction, Replication and Re-enactment in the Humanities and Social Sciences* (Amsterdam: Amsterdam University Press, 2020).

prepare particular ingredients that may be used in other successive operations. Recipes are not ordered randomly, but instead appear in a logical sequence whereby the final product of one prescription often becomes the starting material for the next. As such, the text functions both as a theoretical introduction to the chemical arts, and as a convenient handbook of substances and operations.[48]

However, like the title of the work, the actual arrangement of sections varies across linguistic traditions. This variation makes it difficult to reconstruct the original, Arabic version of the *OAS* – particularly since the only available Arabic manuscript is marred by significant lacunae, the contents of which can be reconstructed only on the basis of the other versions. On the other hand, the Hebrew version appears to be not only complete, but expanded through the addition of recipes that may not have been part of the original text and which are absent in all other witnesses. In order to survey the chemical contents of *OAS*, we must therefore take into account not just the Arabic, but also the Hebrew and Latin versions.

Even within a single linguistic tradition, the ordering of sections can vary considerably. For instance, while the Latin manuscripts and the Hebrew witness all begin with the treatment of vitriols, alums, and salts, before moving on to a section on arsenic, the Latin version published in John of Garland's *Compendium alchemiae* begins with the section on arsenic, then proceeds to the treatment of spirits and metals, while postponing the treatment of vitriols, alums, and salts to the very end of the work. Since Ruska considered the Arabic exemplar of the Garland version highly reliable, he chose to preserve Garland's order first in his own Latin edition, and then in his edition of the Arabic, thus subverting the order of topics actually found in the Arabic *codex unicum* – a decision that I reverse in the present edition.[49]

Fortunately for the success of the *OAS*, the displacement of particular sections does not hinder the general understanding of the work, thanks its organisation into essentially self-standing textual units, each devoted to a specific alchemical ingredient and its preparation. Typically, each of these units describes a particular alchemical substance, and is structured in much the same way. Thus every section starts by describing the physical features of the ingredient in relation to its balance of the four Aristotelian qualities: hot, cold, wet, and dry. The author also introduces bodies (*ağsād*, a common designation for metals) by relating them to their corresponding planets – thus "silver is from the part of the Moon," "gold is from the part of the Sun," and so on. These theoretical remarks are then succeeded by a more or less articulated set of laboratory operations, described in great detail, and with the goal either of preparing the substance for use in other alchemical operations, or of producing gold and silver.

48 On the structure of texts and collections of recipes as criteria of authority, see Chiara Crisciani and Jole Agrimi, "Per una ricerca su *experimentum-experimenta:* riflessione epistemologica e tradizione medica," in *Presenza del lessico greco e latino nelle lingue contemporanee*, eds. Pietro Janni and Innocenzo Mazzini (Macerata: Facoltà di Lettere e filosofia dell'Università di Macerata, 1990): 9–49; Chiara Crisciani, "Fatti, teorie, 'narratio' e i malati a corte. Note su empirismo e medicina nel Tardo Medioevo," *Quaderni di Studi Storici* 108 (2001): 695–717.

49 For a table of correspondences between the sections of the *OAS* in the Arabic, Hebrew, and Latin versions of the *OAS*, see Appendix 3.

Following the order in MS Sprenger 1908, the *OAS* opens with the description of different kinds of vitriol (in modern parlance, metal sulphates). The best type of vitriol is said come from Spain, in particular from the environs of Seville. A section on alums, which are described as greasy substances dried up by earth's dryness, follows, leading to a recipe for the purification of alum using boys' urine. Descriptions of the different varieties of salt and their operations come next, with particular attention paid to alkali salt and sal ammoniac, both of which are defined as the most useful and noblest of salts.

The following sections are devoted to the two kinds of arsenic – yellow and red – and to their purification, sublimation, and washing. The *OAS* also adds two operations for the artificial production of silver that require the combined action of arsenic and cooking salt on copper, iron, or lead. Similar alchemical treatments are given for arsenic and sulphur.

Sulphur is the topic of the following section, containing operations aimed at removing its combustibility and greasiness. The description and treatment of quicksilver occupy the central sections of the *OAS*, which include numerous recipes for its sublimation and congelation.

The focus of the treatise next shifts to the metallic bodies. The writer describes how metals are formed in mines below the earth, using the prevalent mercury-sulphur theory to explain how they develop.[50] General remarks on how to perform operations on metallic bodies are then followed by detailed sections on each individual metal: starting with gold, then moving on to silver, tin, lead, iron, and copper in turn. Each of these sections also includes a number of alchemical operations, ranging in number from three (in the case of tin) to eight (for gold).

The treatise concludes with a section on glass, precious stones, and marcasite that is very unstable in the manuscript tradition. While the Hebrew translation and Arbuthnott's Latin edition preserve this section in its entirety, MS Sprenger 1908 preserves only the portion on glass and precious stones which are, accordingly, the only parts included in Ruska's Latin edition. No trace of this section appears in Steele's Latin edition either, making this element of the *OAS* the most varied and elusive.

The fluidity of this section in the different manuscript traditions can be explained by the fact that this is the only topic to lack an overarching theoretical introduction that could have functioned as a container for these recipes. Moreover, the section's position at the end of the work means that it may have fallen victim to two opposing phenomena: on the one hand, the accumulation of spurious material, often of an

[50] The mercury-sulphur theory derives from the Aristotelian theory of exhalations and was introduced in the Islamic East during the Middle Ages. According to this theory, all metals are constituted by a mixture of two exhalations, mercury and sulphur, in different proportions, degrees of purity, and environmental conditions. See John Norris, "The Mineral Exhalation Theory of Metallogenesis in Pre-Modern Mineral Science," *Ambix* 53 (2006): 43–65; Matteo Martelli, Sébastien Moureau, and Jennifer Rampling, "Theory and Concepts: The Shared Heritage of Byzantine, Arabo-Muslim, and Latin Alchemy," in *A Cultural History of Chemistry in the Middle Ages*, ed. Charles Burnett and Sébastien Moureau (London: Bloomsbury, 2022), 18–23.

operative nature, that tends to appear at the end of treatises in medieval manuscripts; and on the other, the loss of folios that occurs most often at the end of manuscripts. Despite the absence of a theoretical introduction, the *OAS*'s inclusion of recipes for making glass and artificial gemstones reflects a broader interpretation of the field of interest of alchemy, which in its earlier Greek manifestations was not exclusively focused on metallic transmutation, but also on the production of other artificial substances, such as coloured glass and gemstones.[51]

4. The only surviving manuscript in Arabic: MS Sprenger 1908

Only one manuscript witness of the *OAS* in its original language is currently known to survive: the copy found in MS Sprenger 1908 at the Berlin Staatsbibliothek.[52] This manuscript is part of the vast collection accumulated by Aloys Sprenger (1813–1893), the Austrian orientalist and skilled cataloguer, who was also secretary of the Asiatic Society of Calcutta between 1850 and 1857 and later became professor of oriental languages in Bern and Heidelberg (from 1881).[53]

MS Sprenger 1908 is made up of 46 folios of small size (cm. 14×10.5) and, in the absence of a telling colophon, can be roughly dated to the seventeenth century on codicological and palaeographic grounds. It is penned in a very plain handwriting with consistent use of diacritical points and sporadic vocalisation, making it relatively easy to read. With the exception of the first folio, the top left corner of each *recto* page is numbered in pencilled Arabic numerals: this foliation was added to the manuscript at a later stage and may well be contemporary with the modern re-binding of the manuscript and the trimming of the edges of its pages. Catchwords are present on every *verso*, written in the same hand as the main text, but more minute and disposed diagonally. The black ink used to write the headings of new sections has been traced over in red, either by the copyist of the main text or by a later reader.

The manuscript is extensively annotated, containing numerous marginalia of differing length and character. These range from single-word glosses that propose corrections or additions to the text, to more extensive notes that offer entirely alternative recipes to those furnished in the main text. These longer notes, which are found throughout the manuscript, are often penned upside-down in relation to the main text; the minute and uncertain handwriting of these longer notes at times makes them difficult to read, especially given that parts of their text, when

[51] Works on the production of coloured glass and precious stones are included in the Byzantine collections of Greek alchemists and attributed to crucial figures of the alchemical canon, like (Pseudo-)Democritus. See Matteo Martelli, *The* Four Books *of Pseudo-Democritus* ("Sources of Alchemy and Chemistry" 1) (Leeds: Maney Publishing, 2013), 1–69, esp. on 23.

[52] High-quality digital images of MS Sprenger 1908 are available at the STABI Digitale Sammlungen website: https://digital.staatsbibliothek-berlin.de/werkansicht?PPN=PPN892788259&PHYSID=PHYS_0005 (accessed 10 February 2022). The manuscript is listed as number 10361 in Wilhelm Ahlwardt's catalogue and is described in detail in Alfred Siggel's survey of the Arabic alchemical manuscripts in the Staatsbibliothek: Wilhelm Ahlwardt, *Verzeichniss der arabischen Handschriften der Königlichen Bibliothek zu Berlin*, 10 vols. (Berlin: Asher, 1887–1889), VIII, 614–5; Alfred Siggel, *Katalog der arabischen alchemistischen Handschriften Deutschlands: v. I: Handschriften der öffentlichen wissenschaftlichen Bibliothek – früher Staatsbibliothek Berlin* (Berlin: Akademie Verlag 1949), 139–44.

[53] Thomas Schmieder-Jappe, *Die Sammlung der orientalischen Handschriften der Staatsbibliothek zu Berlin* (Berlin: Logos Verlag, 2004), 13.

written too close to the edges of the leaves, was unfortunately lost when the folios were trimmed.[54] In addition, the manuscript includes a peculiar kind of gloss, in which a word from the main text is repeated in the margin in separated letters – that is to say, without the ligature that normally connects Arabic letters. The purpose of these glosses is not obvious.[55]

In the main text, some names of alchemical materials are spelled in the Syriac alphabet or in a mixture of Syriac and Arabic alphabets. This feature is very sporadic and inconsistent: in some cases, the same ingredient is given in both Syriac and in Arabic characters on the same page. Another peculiar feature of the Sprenger manuscript is the use of the system of correspondences between numbers and letters of the Arabic alphabet (known as *abǧad*) for encrypting the names of some ingredients.[56] Two slightly different systems of *abǧad* were in use in the Eastern areas of the Arabo-Islamic world and in North Africa, and I have attempted to derive information on the scribe's origin from such hints. Unfortunately, this attempt has proven fruitless since the only ciphered letters used in the manuscript are those for which the Eastern and Western *abǧad* systems coincide.

4.1. The frontispiece and al-Ṭuġrā'ī

On first glance, readers of MS Sprenger 1908 might be misled into viewing the whole volume as a single work. The frontispiece on fol. 1r includes what looks like a title and statement of authorship: *Kitāb al-ǧawhar al-naḍīr fī ṣināʿat al-iksīr li Abī ʿAbdallāh al-Ṭuġrā'ī* ("*The Book of the Blooming Gem in the Preparation of the Elixir* by Abū ʿAbdallāh al-Ṭuġrā'ī"). A modern, typed note is glued below, amending this title: "On preparing the elixir by ʿAbdallāh Ṭoghráy." Over the page, below the traditional *basmala* ("In the Name of God, the Merciful, the Compassionate") and a lengthy series of invocations and praises to God, another passage again mentions al-Ṭuġrā'ī:

> The poor servant of God, may He be exalted, Muḥammad ibn ʿAbdallāh al-Ṭuġrā'ī, may God be with him, says: I have written this book on the traces and the essence of wisdom; it contains the proof of the reality of the single [components] and their preparation; I have titled it *The Blooming Gem in the Preparation of the Elixir*. In God, the Almighty, I place the hope that this book will be useful for His servants and thanks to Him its message will spread, since He has the power to do what He wants [...].[57]

As already noted by Siggel, the name of the author in the frontispiece is a product of either mistake or invention.[58] There was indeed an alchemist and poet named al-Ṭuġrā'ī, whose full name was Mu'ayyid al-Dīn Abū Ismā'īl al-Ḥusayn ibn ʿAlī

[54] The most extensive notes are at fols. 13v, 14r, 15v, 16v, 17v, 18v, 20r, 21v, 22v, 42v, 43v, and 46v.

[55] Examples of this feature can be seen on fols. 14v, 15r, 16r, 25r, 30r, and 33r.

[56] See Gustav Weil and Georges S. Colin, "Abdjad," in *Encyclopédie de l'Islam* 2, 97–8. The *abjad* system is used very rarely nowadays, mostly when paginating book prefaces and indexes, but was extensively employed throughout the premodern period for the production of astrolabes and talismans, for divinatory purposes, and when composing chronograms.

[57] MS Sprenger 1908, fol. 1v.

[58] Siggel, *Katalog*, 142.

ibn Muḥammad al-Ṭuġrā'ī al-Munshi' al-Īṣbahānī. He was born in Iṣfahān (nowadays in Iran) in 1061 and worked for the Seljuk court, first under Malik Shāh and later as secretary for his son Muḥammad I. He took an active role in the rebellion led by the prince Mas'ūd ibn Muḥammad who, after the defenestration of his brother Maḥmūd III, proclaimed himself sultan and employed al-Ṭuġrā'ī as *wazīr*. However, when Maḥmūd III managed to retrieve his position and regain power, al-Ṭuġrā'ī was accused of apostasy and sentenced to death in 1121 by al-Simīrumī, a *wazīr* of the returning sultan.[59]

This al-Ṭuġrā'ī was also an adherent of alchemy, and the attributed author of several original alchemical works. Among these, the *Kitāb Ḥaqā'iq al-Istishhād* (The Book of the Truths of the Call to Witness), composed in 1112, is notorious for its rebuttal of Ibn Sīnā's philosophical critique of alchemical transmutation. According to Ibn Sīnā, if metallic transmutation were possible it would require the *faṣl* (*differentia specifica*; specific difference) of the transmuted substance to take on the *faṣl* of another metal – a transformation that Ibn Sīnā considered philosophically untenable.[60] Responding to this criticism, al-Ṭuġrā'ī maintained instead that the alchemist's job was to prepare substances to accept a change in *faṣl*; the actual change, however, was effected by God, thus re-establishing the theoretical viability of the alchemical work.

The inclusion of the *Kitāb al-ǧawhar al-naḍīr fī ṣinā'at al-iksīr* among the original works of al-Ṭuġrā'ī appears dubious: Manfred Ullman, who offered a bio-bibliographical portrait of al-Ṭuġrā'ī and an exposition of his defence of alchemy, does not mention this title, and nor does Sezgin, who did not include al-Ṭuġrā'ī among the Arabic alchemists considered in his vast bibliographic work, the *Geschichte des Arabischen Schrifttums*.[61] The only attributions of the *Kitāb al-ǧawhar al-naḍīr* are, as far as I know, found in modern Arabic biographical dictionaries of famous scientists, publications that do not provide any reference to the actual manuscript evidence and therefore are hardly reliable.[62] A survey of the major Arabic collections preserved in European libraries has not revealed any further manuscript by this title that could be connected to al-Ṭuġrā'ī. At the current state of knowledge, I am inclined to consider MS Sprenger 1908 as a compilation of alchemical texts of varied authorship and origin, which was pseudepigraphically transmitted under the name of al-Ṭuġrā'ī only in the manuscript

[59] On al-Ṭuġrā'ī, see François C. De Blois, "Al-Ṭughrā'ī," in *Encyclopédie de l'Islam. Second Edition* (Leiden: Brill, 1960–1996), X, 643–44.

[60] On Ibn Sīnā's opinion on the philosophical impossibility of transmutation and its transmission as Book IV of Aristotle's *Meteorologica* in the Middle Ages, see Chiara Crisciani, "Meteore, IV: 'Sciant artifices alchimie,'" in *Edizioni, traduzioni e tradizioni filosofiche*, ed. Luca Bianchi, Onorato Grassi, and Cecilia Panti (Canterano RM: Gioacchino Onorati, 2017), 353–67.

[61] Manfred Ullmann, *Die Natur- und Geheimwissenschaften im Islam* (Leiden: Brill, 1972), 229–31; 252–3; Fuat Sezgin, *Geschichte des Arabischen Schriftums* (Leiden: Brill, 1971), IV.

[62] See, for example, Maurice Sharbil, *Mawsu'a al-Muktashifīn wa-al-Mukhtari'īn* (Bayrūt: Dār Al-Kutub Al-'Ilmiyya, 1991), 247 [Arabic]; Mustafa al-Jiwasī, *Mawsu'a 'ulamā' al-'arab wa-al-muslimīn wa-a'lāmihim* ('Amman: Dār Asmā' li-nashr wa-al-tawzī,' s.d.), 232 [Arabic]. It is difficult to determine whether these publications are referring to the very same MS Sprenger 1908 or are based on other evidence.

under examination. An overview of the contents of the volume provides further evidence of its composite nature.

4.2. Ms Sprenger 1908, a multiple-text alchemical manuscript

The *OAS* occupies only 17 of the 46 folios of the Sprenger manuscript. An analysis of the material that precedes and follows the *OAS* reveals the kind of materials included in this multiple-text manuscript, their organisation (or lack thereof), and their possible readership.[63]

Even a quick survey of the contents of the Sprenger manuscript shows the lack of any overarching organisation or structure: topics and procedures appear to be juxtaposed without operational or logical order. Only very limited sequences of recipes may be connected by a proximity of ingredients or similarity of procedures and aims:[64] for instance, sections 21–22 focus on animal ingredients; 31–34 on vegetable ingredients and the sublimation of their essences; 40–44 on gold and silver, their cleaning and the transmutation of one into the other; 52–53 on operations on cinnabar; 57–62 on the calcination of metals and stones; 99–100 on the congelation of mercury and its transmutation into silver; and 101–102 on copper. Recipes that take gold and silver as their main ingredients are spread throughout the manuscript, and the same can be said more generally for the recipes dealing with metals.

When compared with the very loose structure of the rest of the Sprenger manuscript, the text of *OAS* stands out for its stricter and more logical organisation, held together by the combination of theoretical passages introducing each set of recipes devoted to a specific ingredient and the recipes themselves. As already mentioned, it is very likely that this orderly progression of topics, together with the fame of its pseudepigraphical author, contributed vastly to the circulation and success of *OAS* in the Latin world.

A further textual element supports the hypothesis of the miscellaneous nature of the Sprenger manuscript and opposes its understanding as a single work by al-Ṭuġrā'ī, as the frontispiece suggests: folios 3r-6r preserve an extensive lexicon of alchemical terminology aimed at recording the *Decknamen* used by alchemists to conceal the ingredients of their operations. The opening of the lexicon clearly states its aim: the author is willing to list all alternative names that the alchemists employ and provide a complete and exhaustive clarification of the baroque accumulation of synonyms used. The lexicon is structurally integrated within a classification of alchemical ingredients into bodies, spirits, and stones, and opens with the following paragraphs dedicated to the names of metals:[65]

[63] For a complete list of the headings and contents of the Arabic manuscript, see Appendix 1.

[64] Section numbers refer to the list of topics presented in Appendix 1.

[65] I have devoted two articles to this lexicon: Gabriele Ferrario, "Understanding the Language of Alchemy: The Medieval Arabic Alchemical Lexicon in Berlin, Staatsbibliothek, MS Sprenger 1908," *Digital Proceedings of the First Annual Lawrence J. Schoenberg Symposium on Manuscript Studies in the Digital Age* 1.1 (2009): Article 2 (online

Description of the names of the seven bodies.

The names of gold: 1 gold, 2 the noble silver, 3 the sun, 4 the father of experience, 5 the jewel, 6 the tomb, 7 the discarded, 8 the knower, 9 the rays, 10 the light, 11 the day, 12 the red wax, 13 the permanent sulphur, 14 the cradle, 15 the balanced, 16 the head, 17 the knot, 18 the integral, 19 the complete, 20 the popular, 21 the patient, 22 the king of the bodies, 23 the gum.

The names of silver: the first[66] the moon, 2 the mother, 3 the leafy, 4 the silver, 5 Hermes of the egg, 6 the white wax, 7 the familiar black, 8 the servant, 9 the night, 10 the leprous gold, 11 the ivory, 12 the path, 13 the adulterer, 14 the braggart, 15 the complete, 16 the ladies' man, 17 the always moving.[67]

The names of iron: the first 1[68] the strong, 2 the inanimate, 3 the forgiveness, 4 Mars, 5 the Indian, 6 the steel, 7 the plug, 8 the rock, 9 the red, 10 the shoot, 11 the distaff, 12 the shredder of the altar, 13 the grave-faced, 14 the rust of the bodies, 15 the filthy servant, 16 the burnt, 17 the upheaval, 18 the misty.

The names of copper: Venus,[69] 1 the skilled, 2 Mars, 3 the airy, 4 the brother, 5 the earth, 6 the green, 7 the yellow, 8 the mediator, 9 the coagulated blood, 10 the lightly colored, 11 the swollen, 12 the fair-complexioned, 13 the generous, 14 the sunny, 15 *Afrūd*.

The names of white lead: *al-qalaʿī*[70], Zeus/Jupiter, 1 lead, 2 the white earth, 3 the soft, 4 [f. 4r] the potter, 5 the white sulphur, 6 the woolshearer, 7 the flabby, 8 the star, 9 Jupiter, 10 lead, 11 the speckled, 12 the white, 13 Jupiter, 14 the mutilated, 15 Zeus/Jupiter,[71] 16 tin, 17 and it is like Jupiter from iron filings and ground myrobalan.

The names of black lead: Saturn, 1 burnt lead, 2 *al-Kāwun*, 3 lead, 4 the earth, 5 the black, 6 the black borax, 7 the smelter of the bodies and the stones, 8 the alkali of the bodies, 9 the smelter, 10 the coal, 11 the body of antimony, 12 the raw/unripe gold, 13 the creek, 14 the Copt, 15 the mint of musk, 16 the dung of the Moon, 17 the reversed gold, 18 the weak, 19 the everlasting, 20 the sulphureous.

The names of mercury: Hermes, 1 the hemiplegic, 2 Mercury, 3 the life of the bodies, 4 the clouds,[72] 5 the horizon, 6 the water, 7 the powerful, 8 the dragger, 8[73] the fastener, 9 the water of the Sun, 10 the writer, 11 the water of the Moon, 12 the water of copper, 13 the water of iron, 14 the composer, 15 the virgin's milk, 16 the girl's milk, 17 the turbid,

65 *Continued*
at: http://repository.upenn.edu/ljsproceedings/vol1/iss1/2); Gabriele Ferrario, "An Arabic Dictionary of Technical Alchemical Terms: MS Sprenger 1908 of the Staatsbibliothek zu Berlin (fols. 3r-6r)," *Ambix* 56 (2009): 36–48. A complete translation and study of this lexicon is due to appear in an anthology of alchemical texts curated by members of the AlchemEast ERC project based at the University of Bologna and edited by Matteo Martelli.

66 While all other numerals in the passage are expressed in figures, this is spelled out.

67 *Siyyār* can also be used for indicating planets on account of their perpetual circular motion.

68 The passage expresses the numeral both in figures and as a word.

69 The numeration of the names for copper begins only with the second of the names listed, leaving out the name for the planet Venus.

70 As above, the numeration does not begin with the first synonym listed.

71 The name *Zāwūsh* is repeated twice in the passage.

72 *ʿanān* (clouds) or *ʿinān* (reins).

73 The number eight is used twice in the passage, thereby causing an error in the numeration.

> 18 the subsistence[74] of all the bodies, 19 the vivifier of the dead, 20 the light of the lights, 21 the sprout of the jinn, 22 the one long in decaying, 23 the soul, 24 the Oriental, 25 the Armenian, 26 the fish, 27 the fugitive, 28 the fleeing servant, 29 the lightning, 30 the heavy water, 31 the humid spirit, 32 the humid body, 33 the water of the snake, 34 the runaway, 35 *al-zīwāqā*, 36 *al-zīwagh*, 37 the water of the lightning of sulphur, 38 the uncovered wine, 39 the water of the glass, 40 the selenite, 41 the foam of the sea, 42 the water of Saturn, 43 the veiled dog, 44 the honey [fol. 4v], 45 the heat of all animals, 46 *da't*, 47 the milk of all the bodies, 48 the medicament, 49 the yeast, 50 the urine of the fools, 51 the sulphur of the two matrasses because it sticks on the lid, 52 *zīwagh* of curdled milk that whitens Venus, 53 work of the ☾[75] without shadow, 54 *zīwagh* of magnesia, 55 *zīwagh* of arsenic, 56 *zīwagh* of sandarac, 57 white lead, 58 antimony of the two explosions, 59 *zīwagh* of sulphur through another and alum, 60 *zīwagh* of Saturn, 61 water of tin and all the bodies and their strength.

The presence of this material in the Sprenger manuscript is interesting not only from the lexicographical and exegetical point of view: a witness of the same text is found among the alchemical writings studied by Berthelot and Duval in the second volume of *La chimie au Moyen Âge*, and, in particular, in MSS Egerton 709 and Or. 1593 of the British Library.[76] Those two manuscripts preserve roughly the same material, but differ in their state of conservation: MS Or. 1593 is very damaged, lacks the opening and closing folios, and some of its quires have been misplaced in the process of re-binding; while MS Egerton 709 is more easily approachable. Folios 1–56 of this manuscript preserve the Syriac text of the *Doctrine of Democritus*, which Berthelot considered to be a product of the period when the practice of extrapolating from and summarising works of ancient authors was diffused in Byzantium. Accordingly, Berthelot describes the first portion of the Egerton manuscript as, "a compilation of alchemical procedures and recipes, translated from Greek around the seventh, eighth, or ninth century [...]. A certain number of these recipes have been passed on almost without any change to the Latin alchemical authors of the thirteenth century."[77]

The second portion of the manuscript (fols. 56–100) is penned by the same copyist, but is mostly made up of Garshūnī material – that is, Arabic language in Syriac script – with the sporadic presence of the Syriac language. This part of the manuscript includes a Garshūnī version of the alchemical lexicon which is found also in MS Sprenger 1908. Berthelot argues that the contents of the second part of the manuscript are later than the *Doctrine of Democritus*, and speculates that

74 Depending on the vocalization, this term can assume slightly different meanings: *Qawām*, straightness, frame, figure, build, strength; *qawwām*, manager, director, caretaker, keeper, custodian, guardian; *qiwwām*, support, stay, basis, foundation, subsistence, livelihood.

75 The half crescent has its "belly" upwards in the manuscript.

76 Berthelot, *La Chimie*, II: I-XII (discussion); 71–74 (edition); 156–61 (translation).

77 Berthelot, *La Chimie*, XII: "une compilation de procédés et recettes alchimiques, traduites du grec vers le VIIe, le VIIIe ou le IXe siècle [...]. Un certain nombre de ces recettes ont passé d'ailleurs, à peu près sans changement, jusque chez les auteurs alchimistes latins du XIIIe siècle." On the alchemical works of Pseudo-Democritus, see Matteo Martelli, *Pseudo-Democrito, Scritti alchemici, con il commentario di Sinesio. Edizione critica del testo greco, traduzione e commento* (Milan-Paris: Arché-SÉHA 2011); Martelli, *The Four Books*.

they could be dated to the period in which the 'Abbasid dynasty and their entourage sponsored a vast movement of translations of Greek medical, scientific and philosophical texts into the Arabic language, often with Syriac as an intermediate passage.[78] Berthelot argues:

> The original Arabic writings are undoubtedly one or two centuries earlier (compared to the Latin translations of Arabic alchemical works): this takes us back to between the ninth and the eleventh centuries for the original redaction of the second part of the Syriac alchemy[79] ... In any event, the composition of this second part is more recent than that of the writings of Christianos, of Anonymous, and of the *Doctrine of Democritus*, but much more ancient than those of the Latin ps-Geber, of Arnold of Villanova, and of the ps-Raymond Lull.[80]

Berthelot's dating hypothesis, cautious and provisional as it may be, provides at least a further element suggesting the miscellaneous nature of MS Sprenger 1908: materials of disparate origin and date were compiled without any apparent ordering principle and with little to no concern about the paternity or the dating of the anthologised texts, but rather with a view only towards the transmission of lexical tools for reading alchemical materials (the lexicon of *Decknamen*), a succinct and practice-oriented theoretical framework (the "theoretical" passages in the *OAS*), and, first and foremost, a large number of detailed descriptions of practical alchemical operations.

4.3. Alchemical authorities, books, and places mentioned in the Sprenger manuscript and in the Arabic *On Alums and Salts*

Throughout the centuries and across languages and geographical, religious, and political boundaries, alchemists have tended to rely on the authority of former masters, who represented – sometimes with the sheer fame of their names and at other times with memorable sentences or specific theoretical contributions – not only an element of continuity for the discipline from its origins through a chain of transmission from master to pupil that validated the discipline itself, but also a pool of textual material that was read, commented, corrected, and repurposed in different times and cultural contexts. Alchemists read the writings of their predecessors in order to unpack the theoretical principles and practical operations they transmitted, in many cases using a secretive language and in a convoluted fashion. The struggle to understand the hidden contents transmitted by

78 On Syriac as an intermediary language in the translations from Greek into Arabic, see Henri Hugonnard-Roche, "Les traductions du grec au syriaque et du syriaque à l'arabe," in *Rencontres de cultures dans la philosophie médiévale. Traductions et traducteurs de l'antiquité tardive au XIVe siècle*, ed. Jacqueline Hamesse and Marta M. Fattori (Louvain-la-Neuve—Cassino: Institut d'études médiévales, 1990), 131–48.

79 This refers to the portion of British Library, MS Egerton 709 that preserves an alchemical lexicon that is also found in MS Sprenger 1908.

80 Berthelot, *La Chimie*, XVII: "Les écrits arabes originaux étaient sans doute antérieurs d'un siècle ou deux [in comparison to the first Latin translations of Arabic alchemical works]: ce qui nous reporterait du IXe au XIe siècle, pour la rédaction originelle des articles de la seconde partie de l'Alchimie syriaque ... En tout cas, la composition de cette seconde partie est plus moderne que celle des écrits du Chrétien, de l'Anonyme et de la *Doctrine de Démocrite*; mais elle est plus ancienne que celle du faux Géber latin, d'Arnaud de Villeneuve et du faux Raymond Lulle."

authoritative alchemical texts is a feature that runs through the whole history of alchemy. This effort was not only devoted to alchemical theories, but involved also more practice-oriented material, like alchemical recipes, which were read, assessed, tried out in the laboratory, and commented upon. Innovations in both theory and practice were introduced in alchemy not so much by the expulsion of a particular past author or text from the alchemical "canon," but rather through the introduction of novel interpretations of authoritative words and works: when alchemists discovered something new, they reinterpreted authoritative texts to accommodate their new findings and re-calibrated their understanding of past ideas and practices in a continuous process of exegesis.[81]

The names of past alchemical authorities are frequently mentioned throughout MS Sprenger 1908 and within the text of the Arabic *OAS*. These citations provide significant elements for reconstructing the intellectual environment of the author, his relationship with specific trends, theories, and practices circulating at the time of composition of the work, and, more generally, the ways in which authoritative texts were used in an alchemical context.

The authority most frequently quoted in the Arabic *OAS* is Ǧābir ibn Ḥayyān, a founding figure of Arabic alchemy, whose life and very existence stand between myth and reality.[82] According to the traditional narrative, which was probably the one available to the author of the *OAS*, Ǧābir was born in the Iranian city of Tūs around the year 721. After his father's death, he started travelling in search of knowledge, both religious and secular, and eventually settled in Kūfa (now in Iraq), where he is said to have established a workshop. According to this narrative, Ǧābir worked for the 'Abbasid caliph Harūn al-Rašīd and was a personal friend of the sixth Imām Ǧa'far al-Ṣādiq. He also had a very strong connection with the

[81] The concept of "practical exegesis" introduced by Jennifer M. Rampling, "Transmuting Sericon: Alchemy as 'Practical Exegesis' in Early Modern England," *Osiris* 29 (2014): 19–34, and subsequently in Rampling, *The Experimental Fire: Inventing English Alchemy, 1400–1700* (Chicago: University of Chicago Press, 2020), is extremely useful for understanding the relationship between authority and innovation in alchemy both in early modern England and more generally.

[82] Ǧābir's name is often transliterated as Jābir in anglophone publications. I use the scientific transliteration of the name Ǧābir throughout the book, but employ the simplified spelling of the adjective "Jabirian," e.g. the Jabirian corpus. The main studies on Ǧābir were conducted by Paul Kraus in the first half of the twentieth century: Paul Kraus, *Jābir ibn Ḥayyān. Textes choisis* (Paris/Al-Qāhira: Maisonneuve/El-Khandgi, 1935); Paul Kraus, *Jābir ibn Ḥayyān. Contribution à l'histoire des idées scientifiques dans l'Islam*, 2 vols. (Al-Qāhira: Imprimerie de l'Institut Français d'Archéologie Orientale,1942–43; repr. Paris: Le Belles Lettres, 1988). Other works attributed to Ǧābir were published in Eric J. Holmyard, *The Arabic Works of Jābir ibn Ḥayyān* (Paris: Geuthner, 1928); Pierre Lory, *Tadbīr al-iksīr al-a'ẓam – L'elaboration de l'elixir supreme* (Damas: Institut Français de Damas, 1998); Pierre Lory, *Jābir ibn Ḥayyān. Dix traités d'alchimie. Le dix premiers traités du 'Livre de soixantedix'* (Paris: Sindbad, 1983). The identity, dating, and the very existence of the alchemist Ǧābir ibn Ḥayyān have been at the centre of a lively scholarly discussion. Kraus maintains that "Ǧābir" should be considered as a collective name for a group of scholars of Shī'ī persuasion active around the ninth century. Although Kraus' hypothesis has been variously criticised – e.g. by Syed Nomanul Haq, *Names, Natures and Things. The Alchemist Jābir ibn Ḥayyān and his Kitāb al-Aḥjār* (New York: Springer, 1994), 3–32, who maintains the existence of a single alchemist called Ǧābir, who flourished in the eighth century and composed all the books that bear his name—it still remains the most plausible. Thijs Delva, "The Abbasid Activist Ḥayyān al-ʿAṭṭār as the Father of Jābir b. Ḥayyān: An Influential Hypothesis Revisited," *Journal of Abbasid Studies* 4 (2017): 35–61, presents important new findings that further problematise the question of Ǧābir's biography and historicity. The most recent and complete discussion of Ǧābir, his works and the status of the Jabirian question is in Regula Forster, "Jābir ibn Ḥayyān," in *Encyclopaedia of Islam* 3, ed. Kate Fleet, Gudrun Krämer, Denis Matringe, John Nawas, and Everett Rowson (Leiden-Boston: Brill, 2017), 91–97.

powerful Iranian family of the Barmakids, prominent and influential viziers to the 'Abbasid caliphs, and suffered their same misfortune when the family fell out of favour at court. Ǧābir then died around the year 815.

Whatever doubts arise concerning Ǧābir's historical existence, he subsequently became established as one of the leading alchemical philosophers in both the Arabic and Latin traditions. This influence is evident throughout MS Sprenger 1908, where Ǧābir is mentioned by name many times, including two references to Jabirian works in the *OAS* itself. First, the writer quotes Ǧābir in a section on arsenic: "Ǧābir said in the [*Book of the*] *Abstractions*" (قال جابر في المجردات, fol. 21v). Here, he quotes the Jabirian work as an authority for the view that arsenic will reach its highest perfection when cooked with bitter almond oil, or another oil.

This "Book of Abstractions" can be identified as the *Kitāb al-Muǧarradāt* (translated as *Liber denudatorum* or *Liber de spoliationibus* in Latin versions of the *OAS*), a work mentioned by the tenth century Iraqi bookdealer and bibliographer al-Nadīm among the works of Ǧābir, and which Kraus listed as the fifty-sixth book of an important Jabirian compilation, the *One Hundred and Twelve Books.*[83] According to Kraus, the inclusion of this treatise among the *One Hundred and Twelve* is doubtful, since its writer expressly claims to have written it after the date of this collection; it is far more extensive than other treatises included in the *One Hundred and Twelve*; and it frequently refers to the *Kutub al-Mawāzīn* (The Books of the Balances), a collection of later production.[84]

Regardless of provenance, the writer of the OAS seems to have turned to Ǧābir for his advice on arsenic in particular. He goes on to quote another work on this subject: "Ǧābir in the [*Book of the*] *Flower Beds* said: 'Arsenic takes the place of the scorpion for the red, while the scorpion takes the place of arsenic for the white.'" The "scorpion" here denotes sulphur; the *OAS* here attributes to Ǧābir the idea that arsenic rather than sulphur should be used in the red work (chrysopoeia), while sulphur should be substituted for arsenic in the white work (argyropoeia). However, the writer goes on to admit, "But I have not tried it."

In the Latin versions of the *OAS*, "The Book of Flower Beds" is rendered as *Tractatus Orti* or as *Liber Urad*, the latter being a simple transliteration of the Arabic word. There are two books attributed to Ǧābir that may be identified with this work: the *Greatest Book of the Flower Beds* and the *Little Book of the Flower Beds*, respectively number 961 and 962 in Kraus's list of Jabirian works.[85] According to Kraus, the longer of these two homonymous works is a technical treatise representing an advanced stage of development in Jabirian alchemy, in which the theory of the balance features heavily.[86]

[83] Al-Nadīm, *Kitāb al-Fihrist*, ed. Gustav L. Flügel, 2 vols. (Leipzig: Vogel, 1871–1872), II, 421; see also the English translation, Al-Nadīm, *The Fihrist of Al-Nadīm. A Tenth Century Survey of Muslim Culture*, ed. Bayard Dodge, 2 vols. (New York/London: Columbia University Press, 1970), II, 857. Paul Kraus, *Jābir ibn Ḥayyān*, 30–31.

[84] Kraus, *Jābir ibn Ḥayyān*, I, 31.

[85] Kraus, *Jābir ibn Ḥayyān*, I, 118–9. A *Kitāb al-riyāḍ* is also mentioned among Ǧābir's works by al-Nadīm, *Fihrist*, II, 423; see Dodge, *The Fihrist*, II, 862.

[86] Jabirian alchemy, at least in the later stages of its development, relies on the assumption that the Arabic letters that constitute the names of substances, as well as the position of these letters in the name itself, carry information on the

While Ǧābir is the most often cited authority, the *OAS* also mentions other past philosophers. Soon after the reference to arsenic and the *Book of the Abstractions* on fol. 21v, the writer goes on to approve the words of a certain Ḫālid), probably to be identified with the Umayyad prince Ḫālid ibn) خالد Yazīd (d. 704), generally considered the first among the Arabs to become interested in alchemy. According to legend, Ḫālid developed an interest in alchemy after dropping his ambition to become caliph.[87] He then ordered the translation of alchemical books from Greek and Coptic into Arabic and held meetings with a monk, named Maryanus or Morienus, who was also an alchemist.[88] The proceedings of these meetings were supposedly recorded in one of the first Arabic alchemical books, the *Risālat Maryanus al-rāhib al-ḥākim li-'l-amīr Ḫālid ibn Yazīd* (Epistle of Morienus, the Monk, the Wise, to the Prince Ḫālid ibn Yazīd). The *Risālat Maryānus* is also considered the first alchemical treatise to enter the Latin world, thanks to the translation that Robert of Chester completed in 1144, which became known under the title *Liber de compositione alchimiae*.[89] In the *OAS*, Ḫālid is mentioned as an authority on mercury, who claims that mercury is characterised by a "secret" property that is absent in other metals, and that allows it to "marry" other bodies. It can be prepared in various ways, including sublimation and repeatedly "killing" it until mercury becomes white. The *OAS* concludes, "Ḫālid was right in his speech." The presence of this quotation among the poems attributed to Ḫālid shows that the author of the *OAS* did not forge the quoted material, but had access to alchemical works; it may also provide some evidence for the dating of the poems themselves.[90]

86 *Continued*
qualitative nature of the substance. The theory of the balance instructs how to calculate the proportions of the elements within a substance and thereby understand which elements should be added or subtracted in order to perfect the said substance. See Kraus, *Jābir ibn Ḥayyān*, II, 187–303.

87 The traditional (and mostly legendary) narrative of the introduction of alchemy to the Arabo-Islamic world and the role of Ḫālid ibn Yazīd in this process is related in al-Nadīm, *Fihrist*, II, 419; see Dodge, *The Fihrist*, 850.

88 The historicity of this narrative has been widely questioned by numerous scholars, since it would imply that translations from Greek into Arabic had already started in the late seventh to early tenth century. The story was strongly criticized by Julius Ruska, *Arabische Alchemisten I. Chālid ibn Jazid ibn Mu'āwija* (Heildelberg: Akten der von-Portheim-Stiftung 6, 1924). See also Ullmann, *Die Natur- und Geheimwissenschaften*, 192–5, and Ullmann, "Ḫālid ibn Yazīd und Alchemie: Eine Legende," *Der Islam* 55 (1978): 181–218, reprinted in *Aufsätze zur arabischen Rezeption der griechischen Medizin und Naturwissenschaften*, ed. Rüdiger Arnzen (Boston/Berlin: De Gruyter, 2016), 334–70.

89 An edition of the Latin translation of the *Risāla* together with an English translation are provided in Lee Stavenhagen, *A Testament of Alchemy: Being the Revelations of Morienus to Khālid ibn Yazid* (Hanover, NH: Brandeis University Press, 1974). For the most up-to-date studies of the *Risālat Maryānus*, including an assessment of the *status quaestionis* on this work, see Marion Dapsens, "De la *Risālat Maryānus* au *De compositione alchimiae*: Quelques réflexions sur la tradition d'un traité d'alchimie," *Studia graeco-arabica* 6 (2016): 121–40. Dapsens' doctoral thesis, defended at the Université de Louvain-la-Neuve in October 2021, includes critical editions of the Arabic text of the *Epistle* and its Latin translation, together with a thorough study of both traditions. Its publication will represent a crucial contribution to our knowledge of the first steps of Arabic alchemy and of the introduction of alchemy to Latin Europe.

90 On the works by Ḫālid, see Marion Dapsens, "The Alchemical Works of Khālid ibn Yazīd b. Mu'āwiya (d. c. 85/704)," *Asiatische Studien/Études Asiatiques* 75 (2021): 327–427. I am very grateful to Marion Dapsens for identifying this quotation in Ḫālid's texts.

The *OAS* often refers to the teaching of anonymous alchemical authorities: the sages (الحكماء, *al-ḥukamā'*; fols. 19v, 20r-v, 27v, 28r), Indian sages (حكماء الهند, *ḥukamā' al-Hind*; fols. 26v, 27v), and the alchemists, who are also called "the people of the work" (أهل الصناعة, *ahl al-ṣinā'a*; fol. 22r). Other quotations are introduced by even vaguer phrases – "others said," "it is said," "he said" – without any hint as to which group or individuals are intended.

Geographical indications are not abundant in the Arabic *OAS,* but some of them are particularly relevant because they featured as important elements of Ruska's critique of the traditional attribution of the *OAS* to al-Rāzī.[91] At fol. 19r, the Persian city of Andara is mentioned as the place of origin of a particular kind of salt (الاندراني, *al-andarānī*). In the same folio, another kind of salt called "bitter salt" is said to be available in "Santa Mariyya" in al-Andalus, in a place called *Balḥūn* (هو بشنتمرية من الأندلس بموضع يعرف ببلحون).[92] Other geographical indications are connected with particular varieties of alchemical ingredients, like Yemenite alum (الشب اليماني , *al-šabb al-Yamānī*; fols. 22v, 23r, 25v, 30r), Cypriot glass (زجاج قبرصي , *zāǧ Qubruṣī*; fol. 23v), Egyptian sal ammoniac (العقاب المصري, *al-'uqāb al-Miṣrī*, Lit. "the Egyptian eagle," "eagle" being one of the *Decknamen* for sal ammoniac; fols. 25v and 29r) and Iraqi glass (الزجاج العراقي , *al-zuǧāǧ al-'Irāqī*; fols. 30r-v).

5. The Hebrew tradition of On Alums and Salts

When novel ideas are translated into different linguistic and cultural contexts, they often call for a new technical vocabulary, resulting in great lexical productivity. In this respect, the trilingual tradition of the *OAS* furnishes a rare source of material for the linguist and lexicographer, demonstrating how technical alchemical terms, often of non-Semitic origin, were received into Arabic, translated into Latin, and rendered into Hebrew.[93]

The Hebrew translation, which is richly annotated and expanded when compared to the Arabic manuscript and the Latin tradition, offers powerful evidence for the reception of alchemical material in a new linguistic milieu. The translation is itself a curious example of multilingualism, in which early modern Hebrew prose is constellated by transliterated Arabic terms and Italian expressions to form a unique text that bears witness to both linguistic struggle and ingenuity. In addition, the translator's notes often suggest alternative procedures or correct the quantities

[91] On the problematic attribution of the *OAS*, see above, § 2.

[92] I have not been able to identify any Spanish place whose name could be connected to the Arabic *Balḥūn*. On the Spanish coast, there is nowadays a locality called Puerto Santa Maria, but I have detected no evidence to connect this site to the one mentioned in the manuscript. The Hebrew translation renders *Balḥūn* with the name *Basilicon*, in transliteration from the Italian language.

[93] On sciences in the medieval Jewish context, see Tzvi Y. Langermann, *The Jews and the Sciences in the Middle Ages* (Aldershot: Ashgate Publishers, 1999); Gad Freudenthal, ed., *Science in Medieval Jewish Cultures* (Cambridge: Cambridge University Press, 2012); Gabriele Ferrario and Maud Kozodoy, "Sciences and Medicine," in *The Cambridge History of Judaism*, Vol. 5: "The Middle Ages," ed. Philip Ackerman-Lieberman (Cambridge: Cambridge University Press, 2021), 825–63. On alchemy in particular, see Gabriele Ferrario, "Alchemy in the Jewish Context," in *A Cultural History of Chemistry*, 87–91.

of the ingredients to be used in a recipe, preserving important traces of reading practices (and possibly of practices *tout court*) in the context of the dissemination of technical alchemical knowledge.

At the same time, the fact that this translation – like the Arabic text examined above – survives in just a single, unique manuscript presents its own difficulties for philologists. Unique manuscripts necessarily limit the amount of material available for analysis by denying the possibility of comparison. Like the tips of icebergs, they may reveal only a small part of their larger, submerged traditions, whose extension and shape must remain a matter for conjecture.[94] Accordingly, my conclusions in this section must rely on an element of speculation. For instance, we cannot know for certain whether the anonymous commentator whose practical knowledge is captured in this translation was the same individual who translated the work into Hebrew, or who later transcribed it. When discussing the Jewish alchemist, I therefore accept that this identity may in fact encompass multiple individuals.

5.1. The Hebrew manuscript Orient. Oct. klein 514 of the Berlin Staatsbibliothek

The Hebrew translation of the *OAS* is preserved in just a single manuscript copy: at fols. 19v-36r of MS Orient. Oct. 514 of Berlin Staatsbibliothek (hereafter MS 514).[95] Written on paper and small in size (cm 14.5 × 10), the manuscript has been dated to the sixteenth century by the Moravian Hebraist and bibliographer Moritz Steinschneider, who describes it as number 258 in his catalogue of the library's holdings.[96] Raphael Patai, who devotes an entire section of his survey of alchemy in the Jewish context to this manuscript, dates it later, to the seventeenth-century. As we shall see, his opinion appears to be more plausible, given that the manuscript includes quotations from authorities who can be identified with seventeenth-century figures.[97]

The whole manuscript is penned by a single copyist in a minute and clear Sephardic handwriting. The scribe employs Quadratic script inconsistently to signal the beginning of new works, new sections within works, or even new recipes. The codex is acephalous, lacking a frontispiece and a colophon, and therefore preserves

94 Paul Maas's opinion resonates loudly for editors of *codices unici*: "Erstausgaben, die auf einem nicht leicht lesbaren codex unicus berühren, bietenselten eine abschließende Entzifferung" ("First editions of texts based on a *codex unicus* not easy to decipher rarely allow for a conclusive transcription"). Paul Maas, *Textkritik*, 2nd ed. (Leipzig: Teubner Verlagsgesellschaft, 1950), 17.

95 High-quality digital images of MS Orient. Oct. 514 are available at the STABI Digitale Sammlungen website: https://digital.staatsbibliothek-berlin.de/werkansicht?PPN=PPN861640624&PHYSID=PHYS_0005 (accessed 17 February 2022).

96 Moritz Steinschneider, *Die Handschriften-Verzeichnisse der Königlichen Bibliothek zu Berlin* (Hildesheim/New York: G. Olms Verlag, 1980; reprint of Berlin: Buchdruckerei der König. Akademie der Wissenschaften, 1878), II, 119–21.

97 Patai, *Jewish Alchemists*, 119–24; 407–16. On the alchemical authorities mentioned in the Hebrew manuscript and their relevance for its dating, see below, 5.7. On translations from the Arabic language conducted in Europe, see Moritz Steinschneider, *Die Europäischen Übersetzungen aus dem Arabischer bis mitte des 17 Jahrhunderts* (Graz: Akademische Druck- u. Verlagsanstalt 1956; repr. of the ed. Wien: Sitzungsberichte Akademie der Wissenschaften in Wien, phil.-hist. Klasse, 101, 1, 1905).

no explicit information on the identity of the copyist or the provenance of the manuscript. It includes two different, co-existing systems of foliation: one using Hebrew letters, the other (and later) one using Arabic numerals. The first extant folio is numbered with the Hebrew letters מ"ב, standing for the number 42, revealing that at some point the first 41 folios of the manuscript were lost, before the second foliation in Arabic numerals was added in modern times. Folios 42–136 in the Hebrew foliation thus became fols. 1–89 in the modern numeration.[98]

Catchwords are found in the lower-left corner of every *verso*: these catchwords run parallel to the main text and are written in the same hand.[99] An interesting note, גויל, appears in the upper margin of fols. 1r-46r, repeated on every page. Steinschneider read this word as a person's name, possibly *gwil*, for Guglielmo.[100] A different interpretation should also be taken into account: *gevil* means "parchment" and also "scroll." The note could therefore refer to the material, and possibly also the format, of the prototype from which this section of the manuscript was translated or copied. A more explicit reference to this exemplar appears on the upper margin of fol. 46v, that reads מ"ס לטינו (*me-s[efer] latynw*, "from a Latin book"); the word לטינו ("Latin") is repeated on the upper margin of the following two pages (47r-v), suggesting that this whole section of the manuscript was translated from a Latin book. Another heading found on only one page (fol. 62v) is of a very different nature: תכליס שמש is a Hebrew transcription from the Arabic تكليس شمش ("calcination of the Sun"), suggesting that it was translated directly from an Arabic source. An Arabic *Vorlage* for this section is confirmed by the presence of Arabic words written in Arabic script within the Hebrew text.[101]

MS 514 is clearly a copy of an earlier manuscript that was perhaps already badly damaged at the time of copying. The scribe's own comments suggest as much. At one point he observes, "I did not find the explanation, because the text was deleted and I wrote what I found" (fol. 17v). The next page includes a similar remark: "of what I have written here, a portion was deleted in the copy of the text because of time, and I have written what I found" (fol. 18r).[102] The manuscript also includes three blank pages (fols. 76v-77v) and several portions of pages are left blank throughout, which we might interpret as representing damaged and unreadable passages in the original manuscript.[103] These pages may have been left blank by the copyist with the intention of filling them in at a later stage, following closer

[98] The discrepancy between the Hebrew and Arabic foliation is explained by a loss of folios. Moreover, the Arabic numeration mistakenly marks fol. 7 as fol. 6, causing a further discrepancy that carries on throughout the remainder of the manuscript.

[99] On styles of catchword in Hebrew manuscripts, see Malachi Beit-Arié, *Hebrew Codicology. Tentative Typology of Technical Practices Employed in Hebrew Dated Medieval Manuscripts* (Jerusalem: The Israel Academy of Sciences and Humanities, 1981; reprint with *addenda* and *corrigenda* of Paris: C.N.R.S., 1977), IV.

[100] Steinschneider, *Die Handschriften*, II, 119.

[101] If this interpretation is valid, we can assume that during the production of this manuscript, the copyist or translator had access to both Arabic and Latin texts. The Hebrew manuscript would therefore become a textual point of confluence of the Arabic (represented by the *On Alums and Salts*) and the Latin alchemical traditions that were circulating together in Europe at the time when this manuscript was produced.

[102] Also noted by Steinschneider, *Die Handschriften*, 119; Patai, *Jewish Alchemists*, 407.

[103] Examples of these blank sections are found at fols. 20r, 21r, 23v, 24v, 25v, 27r-v.

inspection of these problematic passages or the retrieval of a witness preserving a clearer version of the text.

At the current state of knowledge, it is impossible to determine whether the scribe of MS 514 was copying a single, pre-existing Hebrew multiple-text manuscript; bringing together texts from different Hebrew sources; or even serving as translator, by rendering Arabic and Latin material into Hebrew. Each of these options supposes a different level of intervention on the part of the scribe: whether as straightforward copyist, as collector and compiler of multiple manuscripts, or as translator of texts. Regardless of which description best fits our scribe, there is plenty of evidence to suggest that he was actively engaging with the practical content of the texts he recorded – especially in the form of marginal notes and diagrams.

5.2. Evidence for reading practices in MS 514

Although we do not know the identity of the Jewish alchemist who compiled MS 514, we can infer a good deal about his reading practices. The margins of the manuscript are populated by a large number of annotations, penned by the same hand as the main text. These notes are generally introduced by נ"ל, the abbreviation for *nir'eh li*, which can be translated as "it seems to me," or "in my opinion." Similar notes, often enclosed in brackets and introduced either by נ"ל or by ס"א, the abbreviation for *sevarah aḥeret* ("another explanation") also appear frequently within the main text.[104]

Some of these interventions help compensate for missing or incoherent material in the source text. For example, at fol. 5r we find a recipe for congealing mercury that describes how to dissolve alum and work on it until it whitens: it should become greasy but not burn. With this prepared alum, the text explains, one can congeal arsenic and sublimed sulphur.[105] But where is the discussion on congealing mercury that was promised at the start? The copyist/translator adds his own bridge to reconcile the actual text of the recipe with its declared aim. After the aforementioned abbreviation נ"ל ("it seems to me") he notes, "in my opinion, also with mercury and sal ammoniac."

Another folio (fol. 8r) preserves recipes for cerating (or congealing) and melting metals, including some tips for melting mercury. The recipe instructs practitioners to take three parts of mercury that has been previously congealed in copper fumes and two parts of sal ammoniac. Here the reader makes his first intervention, adding – after the ubiquitous abbreviation נ"ל – a single word of comment, "dissolved" (מותך), suggesting that the recipe works better when the sal ammoniac has been previously dissolved. The recipe continues, adding as its final ingredient one part of

[104] For a discussion of these notes in the Hebrew *OAS* and of their significance in regard to the destination of this manuscript, see below, § 5.5.

[105] Congealment generally refers to the solidification of a liquid substance, while here neither arsenic nor sulphur are in a liquid state. While this passage could be simply employing the verb in an extensive meaning for describing the mixing of the two ingredients, the possibility of the corruption of this passage in the manuscript tradition should not be overlooked.

calx of eggshells, which should be added to the other substances before everything is ground together for one day. The product of this preparation is useful for the ceration of metals. The recipe concludes with two further comments by our copyist/translator. The first is introduced by ז״פ (*zeh perush*, "this is a commentary") and suggests dissolving the final product of the operation in order to obtain a water that dissolves all bodies (i.e. metals). This commentary is supplemented by another annotation. A new abbreviation, נ״ל ("in my opinion"), clarifies the process further: the water that dissolves all bodies is obtained, "if you imbibe purified arsenic with the preparation and then congeal it, melt it, and dissolve it."[106]

These are only three examples of the extensive annotations and commentaries that appear on almost every page of MS 514, and which reveal keen and well-informed engagement with the manuscript's technical contents. While in some cases these additions simply integrate portions of text that the commentator overlooked while transcribing or translating his prototype (such as a marginal note on fol. 80r that specifies using a glazed pot for the recipe given in the main text), most of the additions and marginalia suggest practical interventions: clarifying, correcting, and expanding the contents of the main text by supplying additional information on the quantities, details of apparatus, ingredients, and procedures essential to make these recipes "work" in practice. To provide such a wealth of commentary, the copyist/translator indefatigably compared the material in his source texts to his own knowledge. The practical nature of his annotations, which often offer alternative quantities or ingredients to those mentioned in the main text, appears to reflect the lived experience of a practitioner, rather than only knowledge derived from books.

5.3. Drawings of apparatus as "visual annotations" in MS 514

Not all commentary is textual in nature – the copyist of MS 514 also incorporates sketches of alchemical apparatus into his mise-en-page, in order to illustrate and expound the text. The use of visual images is a well-known and widespread aspect of alchemy, with drawings of apparatus constituting by far the most common type, regardless of the cultural context.[107] Depictions of apparatus accompany Greek, Arabic, and Latin texts in manuscript, not to mention European vernacular works. However, due to the relatively small number of extant Hebrew alchemical manuscripts, only a handful of examples of diagrams of apparatus

[106] נ״ל אם תשקה עמו זרניך מתוקו ותשמע ותתיך ותתיר. MS 514, fol. 8r. l. 11.

[107] Most scholarship on visual imagery in alchemical manuscripts has focused on its figurative and allegorical content, especially within the Latin/European tradition: see, for instance, Barbara Obrist, *Les débuts de l'imagerie alchimique (XIVe–XVe siècles)* (Paris: Éditions le Sycomore, 1982). While the Arabic tradition has received less attention, see the important review article, Benjamin C. Hallum, "The *Tome of Images*: an Arabic Compilation of Texts by Zosimos of Panopolis and a Source of the *Turba Philosophorum*," *Ambix* 56 (2009): 76–88. Both Arabic and Latin materials are examined in two recent studies in the same volume: Jennifer M. Rampling, "Art and Representation: The Alchemical Image in the Islamic and Christian Middle Ages," in *A Cultural History of Chemistry*, 149–78; and (specifically on diagrams of apparatus) Nicolas Thomas and Sébastien Moureau, "Laboratories and Technology: Alchemical Equipment in the Middle Ages," in *A Cultural History of Chemistry*, 49–70.

survive in this corpus.[108] MS 514 is therefore important not only as a repository of texts, but also of images.

Like textual annotations, the function of these diagrams can be fully understood only with reference to their anchor text. For instance, on fol. 51r the copyist uses precise, detailed images to illustrate two variations on the same basic equipment (Figure 1). This apparatus appears in a recipe for extracting fat out of egg yolks, which begins on the previous folio.[109] The yolks should be put in a vessel called *ashisha* (אשישה), a kind of cucurbit, and placed over a clay pot. The vessel is then turned upside-down so that its neck fits into a hole previously drilled in the bottom of the pot, and sealed in place. The pot is kept in place by a tripod, so that the neck of the *ashisha* passes through the hole and opens into a receiving vessel below. Four fingers of manure should be placed in the pot so that, when the manure burns, fire surrounds the *ashisha*, causing the fat of the yolks to drip down into the receiver.

The second drawing on this page shows a variation on the same kind of apparatus: when stronger temperatures are required, a second clay pot should be placed upside-down over the first pot, and then sealed. The space around the *ashisha* vessel, between the two pots, is filled with manure and a fire lit on top of the covering pot, so that its heat causes the material inside the *ashisha* to drip into the receiving vessel. In this instance, the translator/copyist, faced with a complex description of apparatus, seems to have used the sketches to clarify how the vessels should be positioned and joined together in order to build this tool for distillation *per descensum*.[110]

Not all the illustrations in MS 514 are so complex, although even simple diagrams can still convey important practical information. At fol. 53r, for instance, the copyist has sketched two crucibles sealed together, the upper crucible having a hole in its base (Figure 2). This kind of double crucible plays an important role in the accompanying recipe. The text directs users to place a black ash, obtained by heating olive oil and vinegar for one night, in the upper crucible, together with some natron and olive oil. The entire apparatus is then placed in a goldsmith's furnace, and the material that drips through the hole into the lower crucible will form a white stone. In this recipe, the configuration of the two crucibles and the hole in the upper one are the key practical features; accordingly, these are the elements emphasized in the diagram.

Another diagram is even simpler: just the outline of a small, round flask, inserted directly into the text of a recipe (Figure 3). The text explains that the flask should be made from either iron or clay with a neck three fingers long, but the drawing does

108 The most important collection of drawings of alchemical apparatus in a Hebrew manuscript is found in the so-called "Gaster manuscript" (British Library, MS Or 10289).

109 The meaning of the Hebrew word שמן describes greasy substances such as fat, grease, and oil.

110 On alchemical operations, see Sébastien Moureau and Nicolas Thomas, "Practice and Experiment: Alchemical Operations in the Middle Ages," in *A Cultural History of Chemistry*, 35–47. On the distillation *per descensum* in particular, see Nicolas Thomas and Caroline Claude, "Les Vases à fond percé: pratique de la distillation *per descensum* au bas Moyen Âge en Île-de-France," *Revue Archéologique d'Île-de-France* 4 (2011): 267–88.

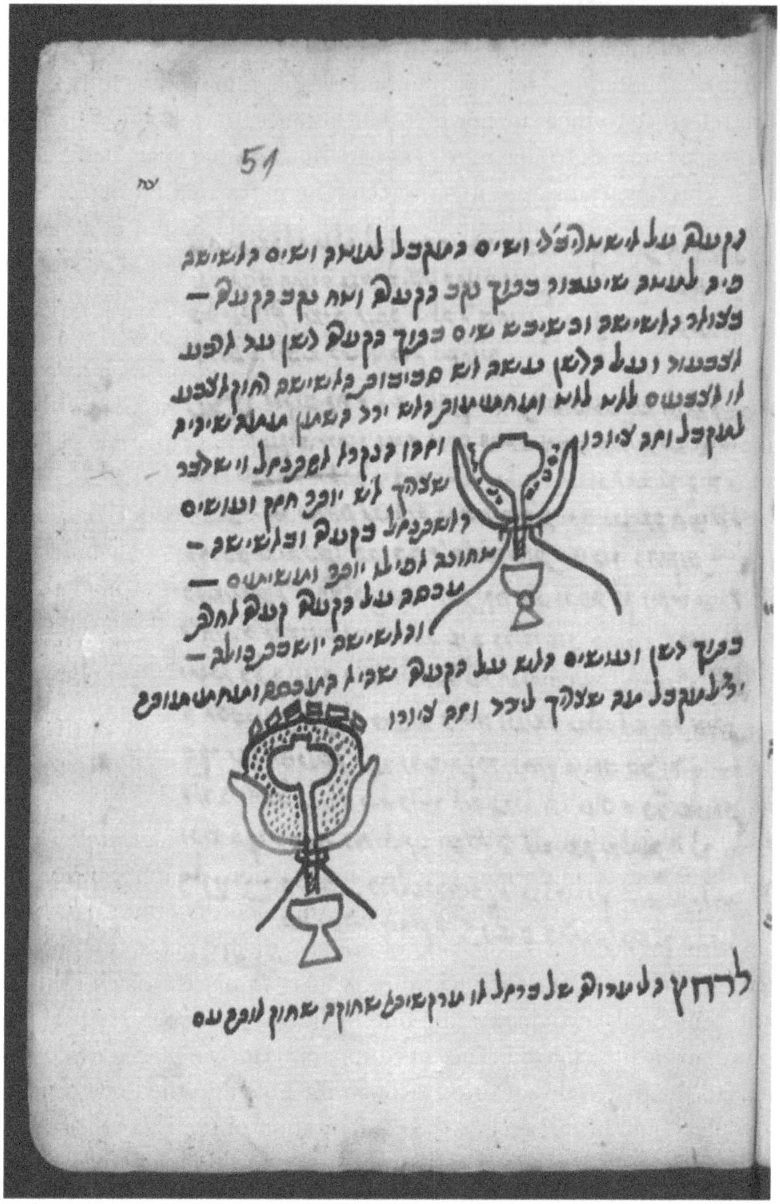

FIGURE 1 Two methods for extracting fat from egg yolks using an *ashisha*. Berlin Staatsbibliothek, MS Orient. Oct. klein 514, fol. 51r.

not convey either the material or the size of the vessel. At first glance, we might wonder why the copyist bothered to sketch such a generic flask. However, since the text does not mention the vessel's shape, the drawing furnishes additional information by suggesting that it should have a round body and an elongated neck – the copyist perhaps drawing on his own knowledge to jot down the appropriate form.

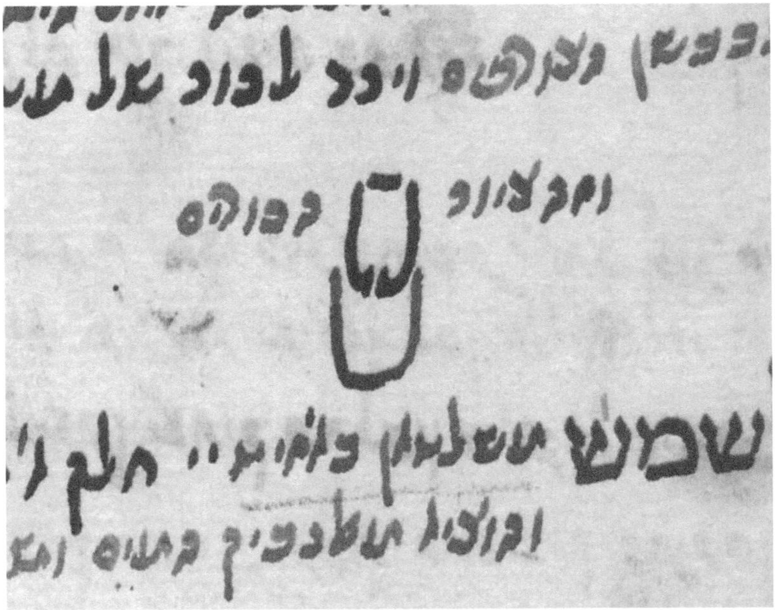

FIGURE 2 Two joined crucibles. Berlin Staatsbibliothek, MS Orient. Oct. klein 514, fol. 53r.

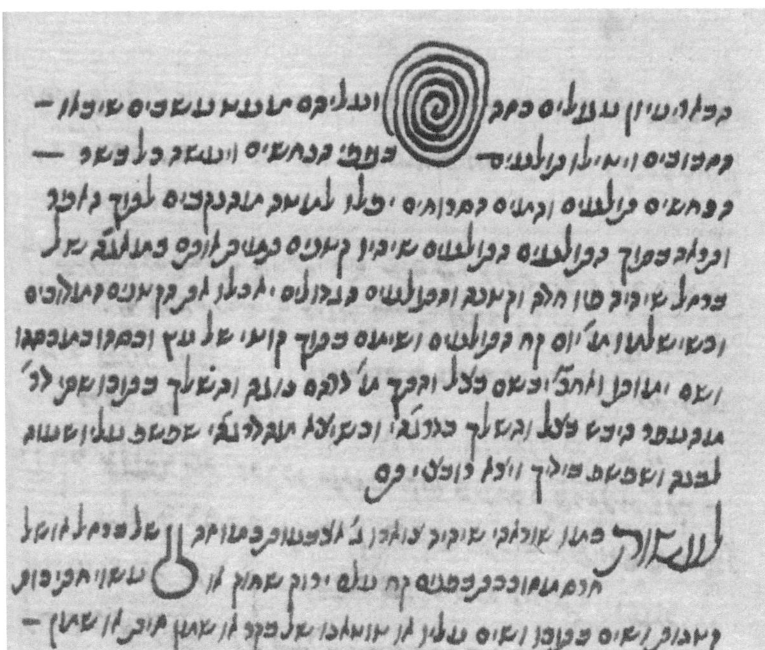

FIGURE 3 Flask and spiral. Berlin Staatsbibliothek, MS Orient. Oct. klein 514, fol. 73v.

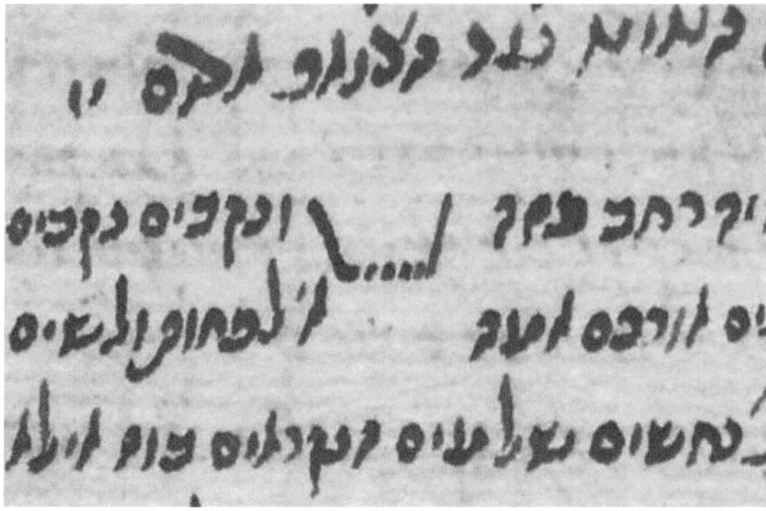

FIGURE 4 Pan with pierced base. Berlin Staatsbibliothek, MS Orient. Oct. klein 514, fol. 73r.

Sometimes, diagrams might help to clarify less conventional apparatus, as on fol. 73r, where a recipe for making a white wax includes some bizarre instructions. The practitioner should start by killing three snakes and pulping their bodies. He should then pound the spleen of a lamb in a mortar and shape it into a thread, before covering it with the pulped snake meat and placing both at the bottom of an instrument described as בארינייון/באריעיון (*b'rynyywn* or *b'ry'ywn*). This term is hard to parse, since the process of transliteration of its non-Hebrew name in Hebrew letters has compromised our ability to decode it. What we can gather from the surrounding text is that the term is describing a pan with a pierced base, the holes covering half of the bottom. This instrument is placed on "three feet," perhaps indicating a tripod. The copyist seems to have recognised the problems presented by both the apparatus and its contents. A sketch of the pan emphasizes the pierced base while omitting the more familiar tripod (Figure 4). He also adds another drawing to illustrate the correct disposition of the snake-and-lamb sausage inside the *b'rynyywn*, to form a spiral shape (Figure 3).

In sum, the Jewish copyist/translator seems to have used drawings of apparatus as a strategy for better understanding his source text, while also providing visual support for his own readers. He treats images as an integral part of his transcriptions, inserting them into the source text much as he does with textual notes – sometimes leaving blank space on the page that he then uses for the drawings; at other times shaping his text around the sketch. He seems to have had no decorative goal in his mind, but we may rather argue that he used drawings as an extension and a clarification of the running text, a form of visual commentary on his source materials that represents his own engagement with their practical contents.

5.4. The place of *On Alums and Salts* in MS 514

So far we have examined evidence for the Jewish practitioner's reading and annotation practices as displayed throughout MS 514, and across a range of texts. These examples provide important context for understanding how the copyist/translator of the Hebrew *OAS* read and used alchemical material, including valuable insight into his views on practice. The structure of the manuscript offers further evidence for how the compiler ordered texts, arranged practical material, and cited authorities.

As we have previously seen in the case of the Arabic *OAS* in MS Sprenger 1908, the *OAS* itself occupies only a small component of MS 514, and appears to be the only extensive section of text offering a logically structured sequence of subject matter. Outside the text of the *OAS*, the manuscript comprises a variety of texts and topics, with little evidence for a coherent organisational scheme.[111] There are a few exceptions where very small sections of the manuscript appear to be arranged thematically: for instance, recipes that are devoted to the whitening of metals and volatile substances (gathered on fols. 5r-6r); the sublimation of metals (ff 7v-8r); and operations for melting (9v–10r). Another free-standing text bearing the title פרקי ביאורים של חכמים ("The Chapters of the Annotations of the Sages," fols. 36r-37r) includes operative passages attributed to Hermes, as well as other authorities who are not mentioned by name, but introduced by the phrase ואמר אחר ("and another one said"). Overall, the contents of the manuscript, together with the abovementioned evidence for Latin and Arabic sources in two of its sections, point to the composite nature of the volume, which appears to be a compilation of alchemical materials generally unified by their eminently practical content.

References to alchemical authorities are abundant in the first 57 folios of the manuscript, and span the whole history of alchemy: from the mythical forefathers of the discipline, such as Hermes, to more recent (possibly even contemporary) alchemists. The manuscript also frequently mentions generalised authorities of the past: "the philosophers" (3v), "the sages" (12v), "the Greeks and ancient Arabs," as well as "some works written in the language of the *goyim*" (48r), and "an Indian" (52r).

The most frequently mentioned name is that of Ǧābir ibn Ḥayyan, who is cited simply as Ǧābir, although some of his attributed works are also quoted by title. For instance, "The Book of the Rectifications of Ǧābir" (ספר תקוני ייבר; *Sefer Tiqqunei Yabir*) is mentioned at fol. 11r and referred to further in the following pages. This title can possibly be identified with the *Kitāb Muṣaḥḥaḥātinā naḥnu* ("The Book of our own Rectifications"), which al-Nadīm placed at the end of the list of *Books of the Rectifications* of ancient philosophers in his survey of Jabirian works.[112]

111 For a complete list of the headings and contents of the Hebrew manuscript, see below, Appendix 2.

112 This work is listed as number 212 in Paul Kraus's survey of the corpus. Kraus, *Contribution*, I, 67. Al-Nadīm, *Fihrist*, II, 422; Dodge, *The Fihrist*, II, 859.

Most of the works attributed to Ǧābir are included in two major collections: the *Book of the Seventy* and the *One Hundred and Twelve Books*. We would therefore expect to find the works listed in MS 514 in one or other of these collections. In fact, for the next three titles quoted in the manuscript, the copyist does indicate their position within the encyclopaedic *Book of the Seventy*. Unfortunately, there is no consistent correspondence between the titles he gives in the manuscript and the list of treatises set down in the actual *Book of the Seventy*. One of these works can be identified as part of the other major collection of Jabirian works, the *One Hundred and Twelve Books*. Similarly, a ס״ הלבנה (*Sefer ha-Levanah*, "The Book of the Moon") is mentioned at fol. 13v, with a note specifying that "this is book number 38 of the *Seventy Books*," although Kraus considers it the forty-fourth treatise of the *One Hundred and Twelve Books*.[113] Folio 14r mentions a ספר ההצלה (*Sefer Ha-Haṣṣalah*, "The Book of the Deliverance") and adds, "which is the twenty-seventh of the *Seventy Books*"; if the title, as it appears, translates the Arabic *Kitāb al-Ḫalās*, this indication in the manuscript matches the position of the treatise in Kraus's study.[114] Folio 16v mentions a ספר מאמר הנחושת (*Sefer Ma'mar ha-Neḥošet*, "The Book Treatise on Copper"), which is described as "the sixty-eighth of the *Seventy Books*"; once again, this indication does not match the order of Jabirian works in Kraus, since there is no treatise by that title in the *Seventy Books*. The work mentioned in MS 514 could possibly correspond to the Jabirian *Kitāb al-Zuhra* ("The Book of Venus"), given the habit of calling metals by the name of the corresponding planet.[115] Further mentions of the name of Ǧābir without reference to any of his works are found at folios 12v and 42r.

Ǧābir is only the most prominent of the authorities cited. Mary the Jew, the mythic initiator of Graeco-Egyptian alchemy and teacher of Zosimus, is mentioned at fol. 13r. In the same folio, we find the name of Aristotle, who is further mentioned at fol. 46v as the author of the ספר סוד הסודות (*Sefer Sod ha-Sodot*, "The Book of the Secret of Secrets"), the notoriously spurious Aristotelian epistle to Alexander the Great. The names of several other authorities are scattered throughout the manuscript: these include *The Book of Archelaos* (ספר ארקילוס, *Sefer 'rqylws*, on fol. 17r);[116] Hermes (הרמיס, *Hrmys*, on fols. 36r, 48r);[117] and Arnold of Villanova (רינאלדו דוילא נובא, *Ryn'ldw dwyl' nwb'*, on fol. 47r). Fol. 48r includes the acronym for Moses Maimonides (רמבם, *RaMBaM*) and the names of Ibn Sinā/Avicenna (בן סינא, *Bn Syn'*); a certain Alberto, probably Albert the Great (אלבירטו, *'lbyrṭw*); Thomas, probably Aquinas (טומאס, *Ṭwm's*); and Roger Bacon (באקון, *B'qwn*).

[113] A *Kitāb al-Qamar* is listed as number 50 in Kraus, *Contribution*, I, 27, where it is referred to as the forty-fourth of the *One Hundred and Twelve Books*. There is a clear discrepancy in the order of titles of Ǧābir's works in Kraus's assessment and in the text of our manuscript. The *Book of the Moon* is also mentioned in al-Nadīm, *Kitāb*, II, 422; see Dodge, *The Fihrist*, II, 861.

[114] A *Kitāb al-Ḫalās* is listed as number 149 in Kraus, *Contribution*, I, 51, where it is referred to as the 27th of the *Seventy Books*.

[115] See Kraus, *Contribution*, I, 74, where the *Kitāb al-Zuhra* is listed as n. 289.

[116] Al-Nadīm, *Kitāb*, II, 419; Dodge, *The Fihrist*, II, 849.

[117] Al-Nadīm, *Kitāb*, II, 417–18; Dodge, *The Fihrist*, II, 845–8.

The mention of a שלטאן כאזיט (*Ślṭ'n K'ziṭ*) at fol. 53r surely indicates the Umayyad Prince Ḫālid ibn Yazīd. However, a reference to a ספר האבנים (*Sefer ha-Avanim*, "The Book of Stones") does not provide enough information to determine which of the numerous lapidaries of Arabic or Latin origin circulating in early modern Europe is intended.

While most of these authorities are standard figures in the medieval Latin tradition of alchemy, the compiler also includes some more specific references that may help us to situate chronologically the production of MS 514 itself, or the exemplar from which its copyist was working. A very intriguing book title, "The Book of the Palms" (ספר אלים, *Sefer Elim*, fols. 48r-v), offers one such clue. Steinschneider did not identify this reference and expressed perplexity on whether "elim" could be the name of a person or a treatise.[118] Patai offered a more intriguing identification: the *Sefer Elim* mentioned here could be the work of that title by the Cretan physician Yosef Šelomoh Delmedigo (1591–1655) that was printed in Amsterdam in 1629 by the Menašeh Ben Yiśra'el. Having left his native island, Delmedigo, also known as Yašar Mi-Qandia ("The Rightful from Crete"), studied mathematics and astronomy in Padua also under Galileo and later devoted himself to medicine under the guidance of Leone of Modena (1571–1648), rabbi of the Venetian Jewish community. In the second part of his life, Delmedigo travelled to Crete, Egypt, and Turkey; he then moved to Vilna, where he worked as a physician, and then to Frankfurt before finally settling in Prague, where he died in 1655.

The *Sefer Elim* collects the correspondence between Yosef Delmedigo and one of his pupil's friends, the Karaite Zeraḥ ben Natan (or ben Nissim) of Troki, in today's Lithuania. In the *Elim*, Delmedigo's answers to Zeraḥ's questions cover a huge range of topics that include astronomy, astrology, qabbalah, alchemy, and occultism. The quotation found in our Hebrew MS consists of Zeraḥ's question about the validity of the Hermetic sciences. As far as the Hebrew *OAS* is concerned, this quotation is important because it sets 1629 as a *terminus post quem* for the production of MS 514.

5.5. The Hebrew translation of *On Alums and Salts*

The Hebrew version of the *OAS* begins at fol. 19v of MS 514 with the incipit, "I begin the Book on Alums and Salts. The treatise on alums and salts that are needed for this operation" (אתחיל ספר אלומי ומלחים. המאמר באלומי ובמלחים הצריכים למלאכה הזאת). The treatise lacks a proper explicit: at 36v, the last recipe of the treatise is followed by a blank space that separates the text of the *OAS* from another text that comes after it.

Comparison between the Arabic and Hebrew manuscripts and the Latin editions of the *OAS* highlights a number of discrepancies in the organisation of the sections and their presence or absence in the different witnesses. In particular, the Hebrew translation – in common with all the Latin witnesses – preserves the sections on

[118] Steinschneider, *Die Handschriften*, II, 121.

mercury and gold and those on talc that are missing from the Arabic manuscript. On the other hand, a number of sections absent from the Hebrew manuscript are preserved either in Arabic, Latin, or both traditions. For instance, the Hebrew lacks the description of one of the operations on tin that is preserved in all other extant witnesses, although not in Steele's edition. The description of one of the operations of iron is absent from all surviving manuscripts except for the Arabic, where it is likely to constitute an addition. The loss of an entire folio in the manuscript is the likely cause of a missing passage in MS 514 devoted to copper, since the Hebrew text stops abruptly at fol. 31v in the middle of a description of one of the operations on iron, and resumes on the next folio in the middle of the description of an operation on copper.[119] Finally, the Hebrew manuscript ends with a section on talc and marcasite that is not found in either the Arabic manuscript or the editions of the Latin translation by Steele and Ruska. At the current state of knowledge, it is difficult to ascertain whether these concluding sections of the treatise should be considered part of the *OAS* or later additions.

Even a quick glance at the Arabic and Hebrew witnesses of the *OAS* reveals notable differences in the length and level of detail of different sections and recipes, with the Hebrew version significantly more extensive than the Arabic. As well as preserving sections that are lost in the Arabic, the Hebrew translation is enriched by a large number of additions in the form of comments, both within the body of the text and in the margins.[120]

As we have seen elsewhere in MS 514, the content of the additions to the *OAS* is eminently practical, often offering suggestions and directions for alternative ingredients and procedures, or corrections in the quantity of the ingredients expressed in the text. I am disposed to believe that the author of these notes was familiar with alchemical practice and wished to correct the recipes in the text, or to supplement them with possible alternatives. It is not always clear whether his comments arise from direct laboratory experience or from extensive bookish knowledge.

In some of these additions, however, the commentator does permit a glimpse of his personal involvement in alchemical practice. Some remarks focus on his own choice of ingredients: for instance, "The [kind of] arsenic that I have tried is the pure crocus-coloured one" (23r). Others recount what appear to be his own attempts to test substances and operations – thus, "I tried alkali salt and I put it on fire and it dissolved quickly, more quickly than any other salt that I have seen" (fol. 21v). Elsewhere, referring to the dissolution of ground sulphur in oil: "I tried it and I dissolved it all" (fol. 26r). An example of a longer marginal note is found at fol. 30v in a description of the treatment of gold. Here the commentator follows the abbreviation for "it appears to me" with suggestions for making the

[119] The foliation in Arabic numerals proceeds without interruption showing that the loss of this folio took place before the figures were added to the margins of the manuscript.

[120] On these additions to the main text, see above, § 5.2.

previous operation on gold more effective: "In my opinion, take an amount of sublimed mercury three times bigger than the amount of gold; take sulphur made white or red in its fire; congeal them in some water of ammoniac salt in which you already dissolved some vitriol. Dissolve everything and let it coagulate. This is a commentary."

Such annotations show that the translator of the Hebrew version of the *OAS* was not merely interested in recording the source text, but also chose to embellish it with a practical commentary of his own. Yet he also sought to situate this experiential knowledge in relation to philosophical opinion and pre-existing authority. "On sulphur," the commentator remarks, "Its nature is hot and humid like the nature of arsenic. Its improvement is like the improvement of arsenic, like the sages said in [their] books regarding this operation. I have not tried this, though. Instead, I have tried sulphur with mercury" (fol. 24v).

The life and practice of this anonymous Jewish alchemist now survive only in partial form, in his marginalia and additions in the Hebrew manuscript, which reveal clear signs of practical interest and knowledge. The authority or perceived antiquity of the *OAS* did not deter him from openly correcting, clarifying, and criticising the text. His additions preserve remarkable evidence for a practice-oriented reading of an alchemical treatise, a work that – we should not forget – was itself eminently operative in outlook.

5.6. The language of the Hebrew *On Alums and Salts*

Can the text of the Hebrew translation of the *OAS* tell us more about its translator, and the context in which it was made? The language of the translation is undoubtedly peculiar, incorporating a mix of languages and even scripts which, taken together, hint at an Italian origin. These peculiarities have long since been noted by scholars – Steinschneider detected the presence of non-Hebrew terminology, and tentatively suggested that it might be Italian.[121] Patai agreed with Steinscheneider, adding that:

> the translator/copyist may have been a good alchemist, but Hebrew was certainly not his best language. He frequently took the easy way out by using Italian (and on a few occasions Arabic) terms in Hebrew transliteration instead of troubling with finding their Hebrew equivalents, and did this occasionally even where perfectly good biblical or talmudic-midrashic Hebrew words were readily available.[122]

Patai also maintains that "the copyist of the extant Hebrew manuscript seems to have had only a rudimentary knowledge of Italian and therefore occasionally distorted the words which in the original Hebrew version were given in Hebrew transliteration."[123]

[121] Steinschneider, *Die Handschriften*, 119: "die Terminologie einer romanischen (italien. ?) Sprache entnommen ist."
[122] Patai, *Jewish Alchemists*, 123.
[123] Patai, *Jewish Alchemists*, 123–24. It is worth noting that Patai considered the copyist of MS 514 to have been working from an existing Hebrew translation, rather than translating from an Arabic *Vorlage*, a view that is still a plausible possibility.

The two scholars assumed that the Hebrew translator was not confident enough with the language of his ancestors and tended to use Italian words for rendering alchemical technical terms that were not available to him in Hebrew. This is only partially true. After completing the edition of the text, I determined that the technical alchemical vocabulary transliterated from Italian constitutes only a part – not even 40% – of the Italian material in this text. In fact, more than 60% of the Italian usage is made up of words of common use and common phrases, as detailed in the tables below.

In these tables, I have transliterated in Latin letters a selection of the Italian words and sentences found in Hebrew (H, first column), their Italian translation (I, second column), their Latin equivalent as found in Steele's edition (L^S, third column), and their Latin equivalent in Ruska's edition (L^R, fourth column). I provide an English translation in the fifth column.

a. Words of common use

H	I	L^S	L^R	Translation
qwnbnynṭw	conveniente	convenientius	convenientius	convenient
byyṭybly	vegetale	plantarum	vegetabilium	vegetable
'wgyynṭw	unguento	oleum	unguentum	grease, oil
qwrynṭw	corrente	currens	currens	running (as in running water)
mynyrly	minerale	mineralis	mineralis	mineral
prsymylyynṣ'	per somilianza	similis	persimilis	very similar
s'ynṣy'	scienza	–	scientia	science
'ysmntw	esaminato	bonum	examinatum	tested, examined
'spyṭzy'wny	*aspettazione*	expectans	expectatio	resistance
s'wpyrnw	si operano	agitur	operantur	are conducted

b. Technical alchemical terminology

E	I	L^S	L^R	Translation
yqwwnṭw	giacinto	–	iaguntia	hyacinth
swlpwrṭh	sulfureo	sulphura	sulphurata	sulphurous
mrqśyṭ'	marcassite	marchasita	markasita	marcasite
fyrmynṭh	fermentata	fermentata	fermentata	fermented
qlṣynmnṭw	calcinamento	calcinatio	calcinamentum	calcination
'slṭsy'wny	esaltazione	sublimatio	exaltatio	sublimation
'wgyynṭw swlwṭybw	unguento solutivo	oleum mollificans	unguentum solutum	grease that dissolves
'lwmy yymny	allume yemenita	alumen jameni	alumen	Yemenite alum
'ynṣyr'	incera	cera	incera	cerate

c. Phrases and sentences

E	I	L^S	L^R	Translation
lsw'h 'wpyrṣy'ny	la sua operazione	modus quidem utendi eius	operatio	its operation
lsw'h qwndymnṭw	il suo condimento	–	suum condimentum	its condiment
dlwqwrpwrly' 'n yspwe'lyṭṭy	dalla corporeità alla spiritualità	a corporeitate in spiritualitate	a corporalitte in spiritualitate	from bodily state to spiritual state
'y qwnpyynṭw 'yrṭyny 'wny qylltrw	e ciascuno trattiene l'altro	et retineat unumquodque aliud	et retinent unusquisque eorum corporem suum	and each of them retains the other
smy'ṭy pry lyqwwly lw pwqw sy mwṭṭ'	e si mette [nei luoghi] in cui si mette il fuoco	in loco ubi incenditur igni	in illis meatibus per quos ignis mittitur	it is placed where the fire is placed
lsw'h 'wpyr'	la sua opera	–	suum opus	its work
byny qwnbwlṭw	ripiegato bene	conquassato	bene convoluto	thoroughly folded
'ylypw 'ysṭrymyyṭṭy d'ysw mylywr'	E la sua *postremità* migliora	laudabilioris successionis	eius postremitates meliores	and its consequences are better
sṭṭw 'dlṭrw sṭṭw	da uno stato all'altro	de dispositione ad dispositionem	de statu ad alium statum	from one state to the other
mlyyr' 'ynsy lsw'h pypryṣ'wny	in essi migliora la sua preparazione	rectificans	meliorat in eis suam praeparationem	its preparation becomes better in them

Even though it can be difficult to distinguish between Latin and Romance languages when transliterated into unvocalised Hebrew,[124] the content of the tables above clearly shows that the words and phrases were originally Italian. This important detail, together with the presence of references to Italy and Italian authors in the Hebrew manuscript, is a notable clue for determining the place of origin of the Hebrew witness of the *OAS*.[125]

In addition, numerous Arabic words are transliterated in the Hebrew text: these words belong consistently to technical alchemical terminology. It is very likely that neither Hebrew technical terminology nor an Italian equivalent was available to the Hebrew translator in these cases, and so he resorted to transliterating Arabic technical terms.[126]

This brings us back to the vexing question of whether the Hebrew *OAS* was originally translated from an Arabic or a Latin prototype. Making such a determination is especially risky given the extent to which medieval Latin translators of Arabic alchemical works depended on and were influenced by the lexicon and

[124] Jean-Pierre Rothschild, "Du Latin à l'Hébreu: quelques problèmes posés par des traductions médiévales," in *Rashi 1040–1990. Hommage à Ephraïm E. Urbach*, ed. Gabrielle Sed-Rajna (Paris: Les Èditions du Cerf, 1993), 697. On the Hebrew translations of philosophical texts, see also Alfred L. Ivry, "Philosophical Translations From the Arabic in Hebrew During the Middle Ages," in *Rencontres de cultures dans la philosophie médiévale. Traductions et traducteurs de l'antiquité tardive au XIVe siècle*, ed. Jacqueline Hamesse and Marta Fattori (Louvain-La-Neuve-Cassino: Institut d'études médiévales, 1990), 167–86.

[125] For the authorities mentioned in the Hebrew *OAS*, see below, § 5.7.

[126] For a list of the words in transliteration from Arabic in the Hebrew *OAS*, see below, Appendix 5.

syntax of their Arabic exemplars. This high level of influence means that any later attempts to render such Latin translations into other languages will inevitably bear deep traces of the language and structure of the Arabic original. Both Steinschneider and Patai engaged with this question and provided diametrically opposed solutions. Steinschneider treated the whole manuscript as a translation from Latin, focusing in particular on the headings that mention a Latin book as *Vorlage* and disregarding the headings that are transliterated from Arabic. Patai instead argued that, at least in regard to the section of the manuscript comprising the *OAS*, the *Vorlage* was Arabic and that "the Hebrew translator used neither the Arabic nor the Latin text published by Ruska, but had before him a different, older Arabic text."[127]

The complete edition of the Hebrew *OAS* now sheds light on this issue, seeming to corroborate Patai's opinion by revealing the very strong presence of Arabic words in Hebrew transliteration. In extant Latin versions of the text, most of these Arabic words have already been translated into Latin. It is also, of course, possible that the compiler of MS 514 – or, more likely, its exemplar – produced the manuscript by juxtaposing several Hebrew versions of different origins, translated from different languages. This hypothesis offers a richer picture of the textual history of the manuscript's contents: one that suggests a confluence of different Hebrew translations of Arabic and Latin alchemical works, already assembled in the earlier manuscript source from which MS 514 was copied.

5.7. Alchemical authorities and sources in the Hebrew *On Alums and Salts*

When and where was the Hebrew translation made? Unfortunately, the unique manuscript copy offers little concrete information on the identity of the Hebrew translator. However, careful study of the text reveals important information about the period and geographical context in which the translation, or its surviving copy, was produced. For instance, the alchemical authorities mentioned in the Hebrew manuscript include a few names that do not appear in the Arabic or Latin traditions, and which offer clues that help us situate the work in space and time.

One recurring authority is introduced by the name of נזרי (*Nzry*). This נזרי is mentioned in two of the additions within the main text that are introduced by the abbreviation נ"ל (*nir'eh li*, "it seems to me"). At 34r, one reads: "נזרי told me that he tried to distil sheets of tin and obtained a white poisonous water." This reference appears to establish a direct transfer of knowledge between the alchemical authority and the writer of the commentary, implying an oral exchange between the two. Shortly before, at 32v, one reads: "נזרי tried this method and roasted [iron] slowly in a hot bread oven: at the sixth time, it melted"; a note that suggests either that the commentator is quoting from *Nzry/Nzr*, or that he has sought to put the authority's advice on roasting iron into practice.

[127] Patai, *Jewish Alchemists*, 120.

One possibility, already suggested by Patai, is that this נזרי may be identified with Giovanni Battista Nazari (fl. 1570), a well-known scholar and alchemist born in Saiano (now Rodengo Saiano), a town located northwest of Brescia, in the wine-producing area known as Franciacorta. Although this identification is nothing more than a conjecture, some elements of Nazari's biography and intellectual production may fit well with the alchemical expertise that our Hebrew commentator attributes to him. Leonardo Cozzando, a contemporary from Rovato (another town in the region of Brescia), provides important information on Nazari's biography and works:

> Giovanni Battista Nazari, with great study and care, managed to show to the eyes of the new city of Brescia the ancient Brescia, its buildings and the placement of its temples. The title is *Brescia Antica* ("Ancient Brescia"), printed more than once in Brescia and finally reprinted by Sabbi in 1658 *in quarto*. In addition he wrote *Della tramutazione metallica sogni tre. Primo della falsa tramutazione sofistica. Secondo della utile tramutazione detta reale usuale. Terzo della divina detta reale filosofica* ("Three Dreams on Metallic Transmutation. The First On False Sophistic Transmutation. The Second On the Useful Transmutation Known as Real Usual. The Third About the Divine One Called Real Philosophical"), printed in Brescia by Ciotti 1599 in quarto, and he also published a *Trattato di Casa Lodrona* ("Treatise on the Lodrona Household"). Moreover, he wrote the *Historia di Brescia* ("History of Brescia"), divided in four parts, and a discourse *De futura, & sperata contra Turcos victoria, e sacris prophetiis, aliisque vaticiniis, prodigis & prognosticis desumptus* ("On the Future and Desirable Victory Against the Turks, Derived From the Sacred Prophets and Other Foretellings, Prodigies and Prognostics").[128]

Vincenzo Peroni, an eighteenth-century historian and biographer of illustrious characters from Brescia, describes Nazari as: "well-versed in letters, in sacred and secular sciences and in antiquities, and a scrupulous collector of native memories. He was well-mannered and, therefore, dear to his contemporaneous literati. He flourished after the middle of the sixteenth century."[129] Nazari is also mentioned by Nicolas Lenglet du Fresnoy, who dismisses him with an unflattering comment in his bibliography of Hermetic philosophy.[130] The name of Nazari also earned mention in a number of alchemically-oriented works published from the sixteenth through the eighteenth centuries: he is cited in

[128] Leonardo Cozzando, *Libraria Bresciana* (Sala Bolognese: A. Forni, 1974; reprint of Brescia: Rizzardi, 1694), 115–16: "Gio. Battista Nazari, con gran studio, e cura, procurò di rappresentar, e porre sotto gli occhi di Brescia nuova la forma di Brescia antica, le sue fabbriche, e giacitura de Templi. Il titolo è *Brescia Antica*, stampato più volte in Brescia, e finalmente ristampato per li Sabbi 1658. in 4. Di più scrisse anco: *Della tramutazione metallica sogni tre. Primo della falsa tramutazione sofistica. Secondo della utile tramutazione detta reale usuale. Terzo della divina detta reale filosofica.* In Brescia Ciotti 1599 in 4. e di più stampò un *Trattato di Casa Lodrona*. Inoltre scrisse l'*Historia di Brescia* divisa in quattro parti, & un discorso *De futura, & sperata contra Turcos victoria, e sacris prophetiis, aliisque vaticiniis, prodigis & prognosticis desumptus.*"

[129] Vincenzo Peroni, *Biblioteca Bresciana*, 3 vols. (Sala Bolognese: A. Forni, 1968; reprint of Brescia, 1816–1823), II, 302: "versato nelle lettere, nelle scienze sacre e profane, nelle antichità, e diligente raccoglitore delle patrie memorie. Fu di maniere soave, e caro perciò ai letterati suoi coevi. Fioriva dopo la metà del secolo XVI."

[130] Nicolas Lenglet du Fresnoy, *Histoire de la philosophie hermétique*, 3 vols. (Paris: Constelier, 1742), I, 313–15; 474: "Jean Baptiste Nazari, Italien, plus grand Compilateur qu'habile Artiste: son livre est assez connu, mais il n'est pas comun."

Evangelista Quattrami's *Vera dichiarazione* ("Truthful Declaration"),[131] in Cesare della Riviera's *Il Mondo magico degli heroi* ("The Magical World of Heroes"),[132] and in Fulvio Gherli's *Proteo metallico* ("Metallic Proteus").[133] Further references to Nazari are found in alchemical anthologies and histories of chemistry that embrace the pre-modern period, including those of Pierre Borel,[134] Johann Friedrich Gmelin,[135] Ferdinand Hoefer,[136] Herman Kopp,[137] John Ferguson,[138] Albert Caillet,[139] Fritz Ferchl,[140] the *Bibliotheca Esoterica*,[141] and Dennis Duveen.[142]

[131] Evangelista Quattrami, *La vera dichiarazione di tutte le metafore, similitudini e enimmi de gl'antichi Filosofi Alchimisti, tanto Caldei e Arabi, come Greci e Latini, usati da loro nella descrittione, e compositione dell'Oro potabili, Elissire della vita, Quinta essenza e Lapis filosofico ... si mostra l'errore e ignoranza (per non dir l'inganno) di tutti gli Alchimisti moderni* (Roma: V. Accolti 1587), 76–77: "Giovan Battista Nazzari nel sogno secondo dice per bocca del Conte di Treues, che il mercurio e solfo de' filosofi sono una cosa istessa uniti insieme, e sono li due semi maschio e femina de' metalli ... "

[132] Cesare della Riviera, *Il mondo magico de gli heroi* (Milan: Arché, 1971; expanded reprint of the 2nd ed., Milan: Pietro Martire Locarni, 1605), 202: "Restino dunque i miseri Alchimisti, poiché così lor piace, erranti, e vagabondi nel confuso labirinto de i loro vani materiali da i quali, dopo la perdita dell'honore, del tempo, e delle proprie facoltà, null'altro raccolgono in guidardone della loro ostinazione, che quelle cinque F, date loro dal Nazari, che dicono: 'Fame, freddo, fetor, fatica e fumo."'

[133] Fulvio Gherli, *Proteo Metalico o sia delle trasformazioni superficiali de'Metalli e delle differenti preparazioni de'medesimi molto proprie per debellare i mali più atroci, che il Corpo Umano affliggono, e per iscoprire gl'inganni de'falsi Chimici* (Venezia: G. Corona, 1721), 154, § "Delle varie trasformazioni del rame." Gherli quotes from a poem found in Nazari's *Della tramutazione metallica*, where the hopes of the alchemists are defined: "frenetiche pazzie, vane Chimere, sogni d'un ebbro, pensier falsi e triti, ladre invenzion ... " On 173–174, § "Delle varie maniere di trasformare il rame in argento," he mentions Nazari's verses featuring the famous five Fs: "apertamente ingannan chi gli crede. / E per l'affaticar che fan gli stolti, / fra tanti alcun non v'è che fe' ritrovi, / perché promette nel principio i mari, / e monti far vedere poi si risolve / in nulla, e per ristor di lor mercede / ritrovan sempre di lor opra al fine / fame, freddo, fetor, fatica e fumo." On 186, in the context of techniques used for whitening copper, he quotes other verses by Nazari: "Non vi beccate tutto il giorno i getti, / né vi formate in cor nuove Chimere, / o privi d'intelletto poveretti / alchimisti d'ingegno, e di sapere. / Fate pur buoni voi vostri concetti / di guadagnar, dovete pur vedere, / che la speranza vana che vi tiene / fa' che perdere 'l tempo, e 'l proprio bene."

[134] Petrus Borellius, *Bibliotheca Chimica, seu Catalogus librorum philosophicorum hermeticorum (mit einem Vorwort von Rudolf Schmitz)* (Hildesheim: G. Olms Verlag, 1969; reprint of Heidelberg: Samuel Broun, 1656), 157, providing some information on Nazari's alchemical works.

[135] Johann Friedrich Gmelin, *Geschichte der Chimie seit dem Wiederaufleben der Wissenschaften bis an das Ende des 18. Jahrhunderts*, 3 vols. (Hildesheim: G. Olms Verlag, 1965; reprint of Göttingen: J.G. Rosenbusch, 1797–1799), I, 299, § 4 "Zeitalter von Paracelsus": "J. Bapt. Nazari, auch als Brescia, mehr als Sammler, als durch eigene bekannt, seine drei Träume und seine *Concordanza dei filosofi* welche gleichsam eine Fortsetzung der ersten ist."

[136] Hoefer, *Histoire de la chimie*, I, 131, § 23: "Alchimistes ambulantes."

[137] Herman Kopp, *Die alchemie von letzten Viertel des 18. Jahrhunderts: die Alchemie in älterer und neuerer Zeit*, 3 vols. (Hildesheim-New York: G. Olms Verlag, 1971; reprint of Heidelberg, 1886), II, 353.

[138] John Ferguson, *Bibliotheca Chemica. A Bibliography of Books on Alchemy, Chemistry and Pharmaceutics*, 2 vols. (Glasgow: J. Macleose, 1906; reprinted in London: Derek Verschoyle Academic and Bibliographical Publications, 1964), II, 131–32. Ferguson reproaches Dufresnoy's judgement on Nazari as simplistic and bibliographically inaccurate and lists works in which Nazari is mentioned.

[139] Albert I. C. Caillet, *Manuel bibliographique des sciences psychiques ou occultes*, 3 vols. (Paris: Lucien Dorbon, 1912–1913, reprinted in Nieukoop: De Graaf Publishers, 1964), III, 168, § 7937.

[140] Fritz Ferchl, *Chemisch-Pharmazeutisches Bio-und Bibliographikon* (Mittenwald: A. Nemayer, 1938), 379 (*non vidi*).

[141] *Bibliotheca Esoterica: catalogue annote et illustre de 6707 ouvrages anciens et modernes, qui traitent des sciences occultes ... , comme aussi des societes secretes ...* (Paris: Dorbon-Ainé, 1940; reprinted in Paris: C. Coulet & A. Faure, 1988), § 3232, where Nazari's alchemical works are mentioned along with a few observations on the illustrations of the 1599 edition of Nazari's *Della tramutazione*.

[142] Dennis I. Duveen, *Bibliotheca Alchemica et Chemica. An Annotated Catalogue of Printed Books on Alchemy, Chemistry and Cognate Subjects* (Utrecht: HES Publishers, 1986; reprint of London: Weil, 1949 with a contribution by H. P. Kraus, "The Duveen Collection of Alchemy and Chemistry"), 426.

Nazari's production, which is still partially in manuscript, reflects his variety of interests: the ancient history of Brescia, the wars against the Turks, the life of the consecrated Angela Merici from Brescia (1470/5–1540), and the bishops of Brescia.[143] His first alchemical work was the *Metamorfosi metallico et humano* ("Human and Metallic Metamorphosis"), printed in Brescia by the Marchetti brothers in 1564. The contents of the *Metamorfosi* constitute the basis for his second alchemical work, *Della tramutazione metallica sogni tre* ("Three Dreams on Metallic Transmutation"), published by the same printers in 1572 and in 1599. The 1599 edition of the *Della tramutazione* is expanded by the addition of a *Concordanza dei Filosofi* ("Concordances of Philosophers") also called *Rosario dei Filosofi* ("Rosary of the Philosophers"). The main theme of the *Della tramutazione* is the description of three highly allegorical dreams representing three different approaches to alchemy, as declared in the frontispiece of the 1599 edition.

Nazari's work also includes a significant catalogue of alchemical authors and their works from antiquity to Nazari's own day that the author claims to have seen inscribed on enormous slabs of marble held by gigantic statues in one of his alchemical "dreams."[144] This catalogue is a precious source on the kind of alchemical bibliography that would have been available to an interested scholar at the turn of the sixteenth century. Unfortunately, the list does not feature the name of any Jewish alchemist who could be identified with the translator/copyist of MS 514, who claims – as we have seen – to have heard some alchemical instructions directly from Nzry. If the identification of *Nzry* with Giovanni Battista Nazari is correct, and if the text is accurate in configuring a direct relationship between the translator/copyist and the Italian erudite, this would provide us with valuable evidence for the place of production and dating of the manuscript. It was possibly composed in Northern Italy (perhaps in the vicinity of Brescia) during Nazari's lifetime or, more precisely, during the lifetime of the Jewish alchemist who met Nazari: that is, between the second half of the sixteenth and the first half of the seventeenth century. If, then, the mention of Delmedigo's *Sefer Elim* is a reliable indication of the *terminus post quem* for its production,[145] we can narrow down the dating of MS 514 to the first half of the seventeenth century. The identity of its copyist remains unknown.

A note on the margin of fol. 29v mentions a certain "Florentine": חלום אמר לי פלורינטינו קח זהב טחון או בלימא או במלוגמא ("Dream. The Florentine told me: take ground gold or filings or an amalgam"). Patai, disregarding the indication that the encounter between our Jewish alchemist and the Florentine may have happened only in a dream, considers this phrase possible evidence for

[143] For a complete list of Nazari's works, see Giovanni Battista Nazari, *Della tramutazione metallica sogni tre* (followed by the *Canzone Alchemica* by Rigino Danielli da Capodistria) (Milan: Arché, 1967; reprint of the 2nd ed., Brescia: Marchetti, 1599, with engravings), III–IV.

[144] Nazari, *Della tramutazione*, 135–44. The same catalogue is reproduced in Lynn Thorndike, *A History of Magic and Experimental Science*, 8 vols. (New York-London: Columbia University Press, 1923–1958), V, 679–95.

[145] See above, § 5.4.

dating the Hebrew manuscript. He suggests two identifications of the Florentine of the manuscript: Florentinus de Valentia, a pseudonym of Daniel Möglin, Rosicrucian and author of an apology for the Rosicrucians titled *Rosa Florescens, contra F.G. Menapi calumnias* (Amsterdam,1617–18); or Antonius de Florentia. A man with the latter name entered the Dominican Order in 1404, became archbishop of Florence in 1446, died in 1459, and was canonised as Sant'Antonino in 1523.[146] However, this second identification appears incompatible with the chronology defined by the proposed identification of נזרי with Nazari: any encounter between the Jewish alchemist who worked on MS 514 and the Dominican Antonius could actually have happened only in a dream. In fact, the list of authorities found in Nazari's *Della tramutazione* does include a mention of one Antonius de Florentia, and of a work titled *Antonii de Abbacia epistola* [*sic*] *due de lapide philosophorum.*[147] The *Bibliotheca chemica* also attributes a work on metallic transmutation to Antonius de Abbatia;[148] in an annotation that relies on Nazari's list, Borel also mentions an Antonius de Abbatia[149] as well as an Antonius de Florentia,[150] although without any indication of dating. Unfortunately, these conjectures do not provide any further information on the identity of the Jewish alchemist who worked on the manuscript of the *OAS*.

Another problematic identification is the one suggested by Patai, who reads the word טסטייני (*tysṭyyny*) as the Italian family name Tastoiani, but without further clarification. The name appears in a marginal note on fol. 31v: חלום טיסטויייני יש בו מלח וחברתי לקליפה ולשמרים. אם כן תוציא החמימות ("Dream *ṭysṭyyny*. In it there is a salt, and I composed it with shell and yeast; if so, the humidity comes out").[151] Patai's identification seems very dubious, particularly since the word *tysṭyyny* appears in other two passages of the Hebrew manuscript: at fol. 13r, where it indicates a substance to be combined with vinegar for the dissolution of arsenic and sulphur, and at fol. 18v in the context of preparing a water that whitens metals.

Despite the conjectural nature of these identifications, it appears that the Hebrew translator/copyist may have had some kind of relationship with alchemists of Italian origin and that his work, if the mention of Nazari and his identification is

[146] Patai, *Jewish Alchemists*, 123.

[147] Nazari, *Della tramutazione*, 140.

[148] Ferguson, *Bibliotheca*, I, 1–2; the work was titled *Epistolae duae scrutatoribus artis chymicae mandatae* and was published both in Latin and in German translations at least four times between 1670 and 1759. The Latin and German text of the *Epistolae* is printed in Friedrich Roth-Scholtz, *Deutsches Theatrum Chemicum*, 3 vols. (Hildesheim-New York: G. Olms Verlag, 1976; reprint of 1732 ed.), III, 651. The identification of the author is difficult. Ferguson proposes to identify him with a monk from Padua, who lived around 1350 and practiced alchemy with some good results; he notes, however, that this identification is incompatible with the presence in Antonius's *Epistolae* of characters who lived in the sixteenth century. Gmelin, *Geschichte*, II, 21, § 1 "Zeitalter der neueren Geschichte, oder Boyle's Zeitalter," suggests that Antonius should be identified with an Italian alchemist contemporary with Robert Boyle (d. 1691).

[149] Borelius, *Bibliotheca Chimica*, 24: "Antonii de Abbacia *Epistolae duae de lapide philosophico*, a Nazari memoratae."

[150] Borelius, *Bibliotheca Chimica*, 24: "Antonius de Florentia, Chimicus est, ex Combachio."

[151] Patai, *Jewish Alchemists*, 123.

valid, could be chronologically placed in the early seventeenth century. The abundant presence of Italian words in transliteration throughout the manuscript, together with the references to Sicily and Calabria (fol. 47v) as the main sources of an unidentified material called *ṭ'rsy'*, appear to point to Italy as the area in which the Hebrew *OAS* took the form that has come down to us.

The journey through the different versions of the *On Alums and Salts* conducted to this point reveals more than the transformation undergone by the text of the treatise in its three linguistic incarnations: it also opens a window on communities of readers, translators, and practitioners from medieval al-Andalus to early modern Italy. *On Alums and Salts* was seamlessly appreciated by readers and practitioners belonging to the Muslim, Christian, and Jewish communities and, similarly to other medieval scientific works, became a point of confluence of and exchange among the three traditions. Muslim, Christian, and Jewish translators and readers left their mark on the extant witnesses of the *OAS*, a mark that speaks more about their shared scientific interest in the doctrines and operations in the treatise than it does about their community of origin and religious affiliation. The doctrinal richness and practical usefulness of this medieval alchemical work produced in a specific cultural context allowed it to be received positively by the other communities and throughout many centuries, in a continuous textual and practical exegetical effort.

EDITIONS AND TRANSLATIONS

Criteria of Edition and Translation

The Arabic and Hebrew traditions of *On Alums and Salts* are represented, at the current state of research, by two *codices unici* and, therefore, the job of the editor is forcefully limited to the respectful representation and tentative interpretation of the available material that constitutes single instances in the history of a complex textual tradition.

In my edition of the Arabic and Hebrew manuscripts of the *OAS*, I have tried to be as faithful as possible to their letter, keeping my interventions to the text as minimal as possible. In the case of *codices unici*, it is not possible to carry out a critical *restitutio* that approximates the archetype of the text on the basis of all the available witnesses. Instead, an edition that respects and reflects the evidence of the surviving manuscript offers, in my opinion, the best entry into a complex text like the *OAS*, and a useful starting point for critical work in case further manuscript evidence should surface in the future.

In my editions, I have decided to preserve the alchemical symbols that are found sporadically and inconsistently in both manuscripts. For their interpretation, the reader should refer to the translation and footnotes. Similarly, in my editions and translations I have not attempted to decode the limited set of *Decknamen* that appear in both the Arabic and Hebrew versions of the *OAS*, but I have instead provided their interpretation in footnotes. I have preserved the division in lines of the two manuscripts and have not attempted to force material that is at times linguistically, grammatically, and syntactically creative – to say the least – into the received standards of medieval Arabic and Hebrew.

For the Arabic edition, I have limited my interventions to the reintegration of missing diacritical dots, in particular the dots under the final ي, that are regularly absent in the manuscript, without signalling them in the apparatus. I have standardised the use of the letter *hamza* to modern conventions; introduced the shortening of the final vowels in forms of the imperative, which is often disregarded in the text; and regularised suffix pronouns when they are clearly the result of a mistake. When in doubt, I have always given precedence to the letter of the manuscript over my hypothetical preferred readings. I have integrated my emendations into the edited text: they are signalled with a Latin letter in superscript that points to the apparatus, which is immediately below the edition of every folio and where the reading of the manuscript can be verified. Rubricated words, that are generally found at the beginning of a new topic or a new section, are rendered in bold font in the edition. Crossed-out words in the manuscript have been reproduced as crossed-out in my edition. Marginal notes that clearly and undoubtedly refer to a particular passage in the main text are positioned in the text itself between asterisks * *; marginal notes that do not refer to specific passages are edited (and translated) at the end of the edition of the folio on which they are found, always between asterisks * *.

In my edition of the Hebrew *OAS*, I have likewise tried to be as faithful as possible to the manuscript and have limited my interventions to the bare minimum required for reading the text. I have preferred to avoid resolving the numerous abbreviations that are found in the manuscript; for their interpretation, the reader should refer to the translation. I have also preserved the only punctuation sign used by the copyist, a single dot (.) that is employed for all kinds of caesurae, from separating the elements of a list to signalling the end of a recipe or of a section. These dots are not used in a consistent way in the manuscript and their irregular use is mirrored in my edition. I have not preserved the straight horizontal lines that the copyist employs to fill up the empty spaces at the end of some lines of text. As with the edition of the Arabic manuscript, I have limited my editorial intervention to a minimalistic approach (almost a diplomatic one). As noted by Patai, the translator or copyist (or both) of the Hebrew manuscript did not master Hebrew style or grammar: I have emended only blatant mistakes that may prevent the understanding of specific passages, while trying to preserve the "rough" style of the Hebrew translation. My emendations are integrated in the main text and signalled by a superscript Latin letter; the apparatus, which is immediately below the edition of every folio, shows the reading of the manuscript. In the edition, I have preserved the overabundant (as compared to standard use) presence of *matres lectionis*, in particular in the imperfect tense of verbs, and the masculine suffix pronoun that in many but not all cases, takes the form נו. Quadratic script is used in the manuscript to highlight the beginning of new sections or new recipes within a section: these phrases are rendered in bold in my edition. Notes and additions are at times enclosed between brackets () in the manuscript: these brackets are preserved in my edition. Marginal notes that clearly and undoubtedly refer to a particular passage in the main text are positioned in the text itself between asterisks * *; marginal notes that do not refer to specific passages are edited at the end of the edition of the folio on which they are found, always between asterisks * *.

In the edition of the Arabic and Hebrew manuscripts, I have introduced section numbers between square brackets [§]: these numbers refer to the corresponding section of the translation, where I have also provided section titles (see below).

The English translations of the Arabic and Hebrew *OAS* face the text of the editions page by page. For the sake of faithfulness to the style of the original Arabic and Hebrew, and in order to allow the reader not versed in those languages to appreciate the stylistic asperities of the two texts, I have always chosen the most literal translation possible: this, in many cases, implies an unfortunate, but scientifically motivated, detriment to the elegance of style that translators of literary material are able and advised to preserve. To allow the easier navigation of the texts, I have introduced progressively numbered section titles between square brackets [ ]; these headings follow the contents of the treatise, but are not found in the manuscripts. These same sections are also used to provide an overview of the material available in the different linguistic traditions in which the *OAS* was transmitted (see Appendix 3). Square brackets [ ] enclose also additions that I decided to introduce for the sake of readability. Rubricated words and phrases in the Arabic manuscript and words and phrases in quadratic script in the Hebrew manuscript are rendered in bold font in the translation. As in the editions, the translation of *marginalia* is enclosed between asterisks * * and, when its positioning could be clearly determined by the presence of pointers in the manuscript, has been integrated in the text; when this has not been possible, the *marginalia* are translated at the end of the text of the folio on which they are found. In the translation of the Hebrew version, the additions and notes that are normally introduced by abbreviations and are often inserted *in medias res* in the main text, are separated from it by em-dashes in my translation (—). The numerous Italian words that are transliterated into Hebrew letters in the Hebrew manuscript are in italics in my translation. Arabic words in Hebrew transliteration are rendered in italics in the translation as well; the original Arabic, at least in the first occurrence of each word, is provided in the footnotes.

In order for the Arabic and Hebrew editions to be displayed facing their English translation, the font size has been reduced in a handful of pages, sacrificing graphical consistency for the sake of clarity.

MS Sprenger 1908 – Berlin Staatsbibliothek

fol. 19r

[...] [§ 1] **صفة استعماله** أن تأخذ من المصوف الأبيض الخالص منه ما شئت فاسحقه ناعما[a] وألقه[b] في قدر مزجج واسكب عليه أربعة أمثاله من أبوال الصبيان بعد طبخ البول وتصفيته ثم حله به على النار اللينة وحركه بعود حرا واتركه حتى يرسب ثفله[c] وصفيه برفق وارفعه لوقت حاجته [§ 2] **القول في الملح** اعلم أن الأملاح كثيرة وأشرفها الأندراني ثم ملح الطعام ثم الهندي [ثم] وهو أحمر وملح الخبز ~~الغير~~ ثم المر وهو المرصوص وهو بشنتمرية من الأندلس بموضع يعرف ببلحون وطبعه حار يابس وهو ماء عقده يبس الأرض ومن خاصيته يذيب القمر ويحرقها لشدة ناريته ويزيدها بياضا ويصرفها في التجسد إلى الروحانية وكذلك يفعل بالشمس ويصيرها حمرة ويغسل الأجساد من الدنس وبه تكلس الأجساد لا بغيره ولبياضه سمته الحكماء ورق الغمامة البيضاء وإن الخلق كلهم محتاجين إليه يصلح أجسام الخلق ويقوم دنياهم وإن الله تعالى لم يمدح شيئا[d] في التورية كمدح الملح وإذا فكرت عرفت فضله ويوجد في جميع أرمدة النبات و الأكلاس والأحجار وعظام الحيوان فالسر كله فيه فمن حله وعقده فقد اطلع على السر المكنون وهو صابون الحكماء وبه أدركت المنى وقررت[e] بأعظم

---

[a] ناعم [b] القيه [c] تفله [d] شيء [e] قرت

MS Sprenger 1908 – Berlin Staatsbibliothek

fol. 19r

[§ 1. On alums]

Description of its use.[152] Take as much of the woolly,[153] white, purified [substance] as you like. Crush it finely, throw it in a glazed pot and pour over it four times its amount of boys' urine, after the urine has been cooked and filtered. Then dissolve it with it on a gentle fire, mix it with a stick while it is hot and let it rest until its sediment settles down. Filter it gently and preserve for the time of need.

[§ 2. On salts]

Discourse on Salt. Know that there are many [kinds] of salts and [that] the noblest among them is the salt of Andara,[154] then there is kitchen salt, then Indian salt[155] which is red, bread salt and then the bitter one which is compact and is [found] in Šantamariyya in al-Andalus, in a place known as Balḥūn.[156] Its nature is hot and dry. It is a water that the dryness of the earth coagulated. Among its features there is [the fact that] it melts the Moon, it burns it thanks to the strength of its fieriness, it increases its whiteness and makes it change from corporality to spirituality. In the same way it acts with the Sun: it makes it become red and washes the bodies from impurity. The bodies are calcined by it and by nothing else. Because of its whiteness, the sages called it the leaf of the white cloud.[157] Every creature needs it: it is beneficial to the bodies of creatures and rectifies their [life in this] world. And God, may He be exalted, did not praise anything in His Law as much as He praised salt. And if you think [about salt], you should know its excellence. [Salt] is found in all vegetable ashes and in calx, in the stones and in animal bones. The whole secret is in it and whoever has dissolved it and has coagulated it, he has attained the hidden secret. It is the soap of the sages, and thanks to it you have reached what is intended and obtained the best

[152] The Arabic *OAS* does not preserve the theoretical passages on vitriol that are found in the Latin tradition. Instead, it opens abruptly with the recipe given here.

[153] *Muṣawwaf* ("woolly") refers to a particular feature of Yemenite alum. Steele's Latin edition (p. 15, § 4) may preserve a more complete reading: "de alumine Jameni albo lanoso puro," while Ruska does not include a corresponding Latin adjective in his edition. The Hebrew version here uses an Italian word in transliteration: *luminoso* ("shiny"). The proximity of the Latin/Italian terms *luminoso/laminoso/lanoso* should also be taken in consideration.

[154] Andara can be identified with a town located in the Kerman province in modern day Iran.

[155] The copyist inserted a ثم here, which I consider a copying error and avoid in translation.

[156] For geographical names in the Arabic *OAS*, see *Introduction*, 4.3.

[157] Ruska, *Das Buch*, 122, proposes to emend الغمامة to the more common العامة and translates this phrase as *silver of the people* ("Silber der Menge"). In his edition he seems to overlook the presence of the adjective البيضاء in the manuscript. In my edition and translation, I chose to follow the reading of the manuscript.

fol. 19v

الأشياء وهو عند العالم يسير وعند الجاهل عسير وإنما هلك طالب هذا الشأن[a] بجهله في التدبير الذي كتموه وأخفوه وملؤوا[b] منه الكتب ولم يصرحوا به أقسم بالله أن رجلا لا يعرف سر الملح لا يدرك قليلا ولا كثيرا مما يحل له الله[c] وإنه إذا جهل سره كرجل يرمي عن قوس لا وتر فيه [§ 3] صفة **استعماله** أن تأخذ منه جزأ[d] فاسحقه ناعما بالماء وضعه في قدرة وغطها[e] بغطائها[f] ويكون مشدود الوصل واتركه في فرن الخبز ليلة ويوما[g] ثم حله من الماء العذب[h] في مثله ثم عقده يأتي مثل الثلج فارفعه واستعمله محلول إلا لم تنتفع به شيئا [§ 4] **القول في ملح القلي** اعلم أن ملح القلي هو أجود الأملاح وأحسنها وأنسبها للمزاج بملح العجين فإنه يعقده ويثبته ولذلك يسمى صاحب الأحجار المثبت وهو قائم[i] بنفسه وبغيره إنما صار أشد الأملاح في القوة لأنه نباتي والنباتي له قوتان قوة المعدن وقوة النبات ويفضله ملح الحيوان بقوة ثالثة وهي الحيوانية وتأمل خاصية ملح القلي بالجندل الأبيض كيف يصفيه ويرده زجاجا أبيضا[j] وليس فيه بشيء[k] من الأملاح وهذه خاصيته[l] ولعمري لقد جربنا ملح القلي على النار فرأينا ملحه أسرع ذوابا من الأملاح كلها بكثير لأن فيه دهنية بيضاء غير محترقة وبه يستغنى عن غيره وطبعه الحرارة والرطوبة لأنه دهن عقده يبس النار *وقيل طبعه حار يابس وهو ثابت للنار * يغلب كل طائر[m]

---

[a] الشان [b] وملؤا [c] اكله [d] جزوا [e] غطيها [f] بغطايها [g] يوم [h] العدب [i] قايم [j] ابيض [k] شيء [l] خاصية [m] طاير

fol. 19v

things. This is easy for the sage, but difficult for the ignorant. Indeed, the seeker of this condition perishes because of his ignorance of the operation that [the sages] hid, concealed, and with which they filled the books, but did not explain. I swear to God that a man who does not know the secret of salt does not obtain either little or much of what God allows him, and by ignoring its secret, is like a man who would shoot with a bow without a bowstring.

[§ 3. Preparation of sal ammoniac]
Description of its use. You should take a part [of salt], grind it finely with water, place it in a pot and cover it with its lid with a firm join. Leave it in a bread oven for one night and one day. Then dissolve it in the same amount of fresh water and then congeal it: it will become like snow. Take it up and use it after dissolving it, otherwise you cannot not make any use of it.

[§ 4. On alkali salt]
Discourse on Alkali Salt. Know that alkali salt is the most excellent of the salts, the most beautiful and the most suitable for mixing with the dough salt, since it coagulates it and fixes it. For this reason, it is called the lord of the stones and fixer. It exists by itself and [can be obtained] by other than itself,[158] thus it is the strongest of the salts, as far as faculties are concerned, because it is vegetable and vegetable [things] have two faculties: the mineral faculty and the vegetative faculty. The animal salt overcomes it thanks to a third faculty, the animal faculty. Consider the specific property of alkali salt with the white river stone:[159] how it purifies it and it turns it into white glass, and there is nothing of salts in it: this is its specific property.[160] I swear on my life that we tried alkali salt on fire and we saw that the melting of this salt is by far quicker than that of any other salt, because it contains an unburnt white oiliness: thanks to it there is no need of anything else.[161] Its nature is hotness and humidity, because it is an oil that the dryness of the fire coagulated. *It is said that its nature is hot and dry and that it is stable in the fire*[162] and it subdues any volatile [substance]

[158] This expression appears to echo the wording of medieval Islamic philosophical debates on the different kinds of existence that define God as the only "existent by itself" and everything else as "existing by other than itself." However, its usage in the above passage suggests a more trivial understanding: alkali salt can be found in nature, but can also be produced artificially.

[159] The "white river stone" possibly indicates flint, a stone used in glassmaking. Flint combined with alkali salt turns into a white-coloured glass when heated.

[160] The Arabic text appears problematic in this passage, possibly due to errors in the copy. My translation is, therefore, tentative and provisional here.

[161] Literally, of "its salt." This passage underlines the peculiar behaviour of alkali salt exposed to fire: its quickness in melting is caused by the greasiness that is contained in it. According to the *OAS*, this quality makes it more suitable to be employed in alchemical operations, since it can be melted rather than dissolved in another substance.

[162] The passage between asterisks appears as a marginal note. A graphic pointer in the text shows the exact position where the copyist intended this note to be positioned.

fol. 20r[163]

من المعدنية كالعبد والعقرب والعلم ويعقدها ويمسكها في النار ويحل الأجساد كلها فاستغنى به عن غيره كما استغنى الزجاج به عن غيره لأنه يكلس الجندل ويحله ويذيبه فكذلك لا شك يكون عمله بالأجساد والأرواح فاعتمد عليه ففيه ما ذكرته الأوائل [a] فكان منه الأسرار العجيبة والأعمال الجليلة فقد كشف لك والله المكنون [§ 5] **صفة استعماله** أن تأخذ من شب العصفر جزأ [b] واسحقه جيدا [c] وتجعله في قدر نحاس أو قدر مريخ وألق عليه ستة [d] أمثاله من الماء العذب واطبخه على النار حتى يذهب من الماء السدس واتركه حتى يرسب ثقله فخذ صافيه واحمله في قدر نحاس أو قدر مريخ واطبخه على النار برفق قليلا قليلا حتى ينعقد مثل البلور ملح شريف ارفعه واحتفظ به [§ 6] **القول في صنعة التنكار** خذ منه جزأ [e] فحله في مثليه من ماء الجبن بعد تعتقه أياما وخذ من ملح الخبز مثله فحله في مثليه من الماء العذب وقد حللت في الماء العذب واطبخه على نار لينة قليلا قليلا حتى ينعقد مثل المهى [f] فارفعه واكتمه فهو من فوائد [g] الجليلة إذا أردت أن تذيب جسدا وحجرا فإن كان جسدا فخذه مهبا أو مكلسا [h] أو برادة وإن كان حجر مسحوقا غبارا فاسحقه به سحقا ناعما [i]

*وقيل طبعه حار يابس وهو ثابت في النار *

[a] الاوايل [b] جزو [c] جيد [d] ستت [e] جزوا [f] المهي [g] فوايد [h] مكلس [i] سحق ناعم

[163] The upper margin of the page preserves an extensive note that is difficult to read, written perpendicularly to the main text and penned by a fairly hasty and less accurate hand than that of the copyist.

fol. 20r

among minerals, like the servant[164] and the scorpion[165] and the sign.[166] It coagulates them, and holds them in the fire, and dissolves all the bodies, so there is no need of anything else apart from it, in the same way that glass does not need anything else apart from it, because it calcines, dissolves, and melts the river stone. Therefore, there is no doubt of its action on the bodies and the spirits. Rely on it, since it contains what the ancients mentioned, and the wonderful secrets and the beautiful actions come from it. This has been revealed to you and God is the One who is Hidden.

[§ 5. Preparation of alkali salt]
Description of its use. You should take one part of safflower alum,[167] grind it finely and put it in a pot of copper or in a pot of Mars.[168] Throw on it six times its amount of fresh water and cook it on the fire until one-sixth of the water evaporates. Let it rest until its sediment has sunk to the bottom. Then take its purified [part] and put it in a pot of copper or in a pot of Mars. Cook it on the fire gently, little by little, until a noble salt coagulates like crystal. Take it up and save it.

[§ 6. Preparation of borax]
Chapter on the art of borax. Take one part of it and dissolve it in two parts of whey that you have already aged for a few days. Take the same amount of bread salt and dissolve it in twice as much fresh water – and it should already be dissolved in fresh water.[169] Cook it on a gentle fire little by little until it congeals like crystal. So, take it up and hide it because it is one of the most useful things. If you want to melt a body or a stone, if it is a body, then take it powdered or calcined or in filings, and if it is a stone [take it] finely ground, and then grind it to a fine powder

164 "Servant" is a common *Deckname* for mercury, which is also commonly called "the fleeing servant" in virtue of mercury's volatility, which causes it to flee the fire.

165 "Scorpion" is a common *Deckname* for sulphur.

166 "Sign" or "Mark" is a common *Deckname* for arsenic.

167 The Arabic expression literally translates as "alum of safflower," possibly a reference to the colour of the alum.

168 Mars is a common *Deckname* for iron.

169 This passage is problematic, since it implies that one should dissolve in water something that has already been dissolved in water. Ruska, *Das Buch*, 53 proposes to emend the second عذب ماء ("fresh water") in the sentence as عسل "honey" on the basis of the Latin versions. The passage should therefore read: "dissolve it in two parts of fresh water in which you have already dissolved honey." This is a very good example of how change in the reading of a single word (or of even just few letters) determines a drastic alteration of the underlying chemistry.

fol. 20v

وأوضعه في بودقة مزججة [a] مطينة الفم وانفخ عليها في النار حتى يذوب مثل الشحم إن شاء الله واكتم هذا الملح الشريف واحمد الله على ما أولاك فأنه عجيب وبه ينعقد العبد وتقام الأجساد وتقيد الأرواح وهو الذي رمزته الحكماء وكتمته وهو نشادرهم وخلهم وملحهم وتنكارهم فاعلمه [§ 7] **القول في العقاب** اعلم أن العقاب أفضل الأملاح وأشرفها في التدبير فإنه يحل العبد ويرده ماء جاريا إذا صعد معه وأدخل الحل في موضع ندى وهو دهن عقده يبس النار وطبعه حار ويابس لطيف غواص نافذ [b] وهو روح طيار معين الإكسير ولولاه ما تم إكسير ولا أنحل ولا تداخل وفيه كمن السر وهو الخراج الولاج المشمع لجميع الأجساد ويخرج السواد من عمقها ويحللها وهو الحجر الفاعل المؤلف بين الأضداد جامع الأرواح مع الأجساد وهو الحجر الذي يشبح فيذهب ويبقى وله إثر الشبح ما بقي الدهر وفيه السر أي ألحمه إذا حللته وحللت ملح العجين ومزجتهما وعقدتهما حتى يأتيا كالجليد قد امتزجا وخلدا وأمسك كل واحد منهما صاحبه وثبتا في النار ولم يتدخنا فطاعم منهما كل جسد تريد حله فإنه يذيب ويحله بإذن الله ولا سيما الحديد واعلم أن كل وافقه في تدبيره شيء مما

---

[a] مزجج [b] نافد

fol. 20v

and place it in a glazed crucible, the mouth of which is sealed with clay. Blow on it in the fire until it melts like fat, God willing. Hide this noble salt and praise God for what He has entrusted to you. Indeed, it is something wonderful. With it the servant coagulates, the bodies are rectified, and the spirits are guided; this is what the sages alluded to and what they concealed, and it is their sal ammoniac, their vinegar, their salt, and their borax. So, know it!

[§ 7. On sal ammoniac]
Chapter on the eagle.[170] Know that the eagle is the best of salts and the noblest for the preparation, since it dissolves the servant and reverts it into running water if it is sublimed with it and if the dissolution is carried out in a moist place. It is an oil that the dryness of fire has coagulated. Its nature is hot and dry, thin, piercing, and penetrating. It is a volatile spirit prescribed for the elixir. Were it not for the eagle, no elixir could be completed and nothing would be dissolved nor penetrate. The secret is hidden in it: it is what exits and enters and cerates all the bodies, extracts the blackness from their depths, and makes them melt. It is the active stone, the combiner of the opposites and the unifier of the spirits with the bodies. It is the stone that extends, then vanishes and remains, leaving a trace of the extension that does not remain forever. In it there is the secret, namely: join it once you have dissolved it, have dissolved the dough salt,[171] mixed them, and congealed them until they became like ice: they are mixed and united and each of them is attached to its companion; they become stable in the fire and do not fume. Feed with them every body you wish to dissolve, it will melt it and dissolve it, with God's permission, and iron in particular. Know that whoever agrees that there is in its operation something

[170] Eagle is a common *Deckname* for sal ammoniac.
[171] Probably a reference to bread salt (see § 2 "On salts"), although in this instance the Arabic text does specify dough rather than bread. I have maintained the distinction between "bread salt" and "salt of dough" in my translation despite the two phrases appearing to refer to the same kind of salt.

fol. 21r

مدحه الله فقد فاز وأصل النشادر من الزبول والرطبات الحيوانية ويتولد من مستوقد الحمامات في قعر القدر والأزقة وبحكمة أهل الصنعة بالتصعيد فلذلك صار سيد الأملاح و وأجلها وليس فيها روح غيره ولأنه في الاستحالة الثالثة [§ 8] **صفة استعماله** أن تجعله مصران في قدر في زجاجة ويفرش له ورق الكرفس أو ورق السلق الأخضر والأحسن أن تصعده ثلث مرار إن أردت تصرفه في الرصاصين والنحاسين والمريخ وإن صرفته في الشمس والقمر فصعده ٣ مرار فإنه ‡ [§ 9] صفة **ذلك** تأخذ من العقاب جزأ [a] فاسحقه بمثله ملحا وصعده في أثال الملح ٣ مرات [b] او ٧ وجدد له الملح في كل مرة تجده مثل الثلج في الترس إن شاء الله تعالى صدق في قوله جميعه [§ 10] **القول في الأنفس** والأرواح المعدنية و أولها العلم والعلم نفس وهو معدن وطبعه الحرارة والرطوبة وهو على ضربين أحمر وأصفر والذي جربته الأحمر المصفى [c] إذ ليس فيه منه [...] ولا أشد إحراقا لجميع الأجساد والعلم يشبه العقرب في كثير من الأعمال وفي سرعته الذوب وقلة [d] الصبر على النار والأصفر أصبر على النار من الأحمر والأصفر أغزر صبغا [d] الأحمر غاية ومن خواصه يبيض الزهرة إذا صعد ويسود

---

[a] جزوا [b] مراة [c] المصفا [d] قلت [e] صبغ

fol. 21r

that God praised will be successful. The origin of sal ammoniac is in the urines and animal humidities, and it is generated in the bath-heaters, at the bottom of the cooking pot and of the chimneys,[172] and by the wisdom of the People of the Work through sublimation. For this reason, it became the lord of salts and the most important of them, and among them there is no spirit other than it. And this is also because the third transformation [takes place] through it.[173]

[§ 8. Preparation of sal ammoniac (i)]
Description of its use. You should place it in a piece of gut in a pot [or] in a flask and it should be covered with a leaf of celery or a leaf of green chard. It is best to sublime it three times, if you want to transmute it into the two leads,[174] the two coppers,[175] and Mars. If you transmute it into the Sun and the Moon, then sublime it three times and it will be ‡.[176]

[§ 9. Preparation of sal ammoniac (ii)]
Description of this [operation]. You should take one part of the eagle,[177] grind it with the same amount of salt, and sublime it in the aludel of the salt three or seven times. Add fresh salt to it every time you find it like snow on the disk,[178] if God – may He who is truthful in everything He said be exalted – wants.

[§ 10. On arsenic]
Discourse on the souls and the mineral spirits. The first of them is the sign. The sign is a soul and a mineral and its nature is hotness and humidity. There are two kinds of arsenic: the red one and the yellow one.[179] The one that I have tried is the purified red one, since there is not in it [...][180] than it nor stronger in burning all the bodies. The sign is similar to the scorpion[181] in many operations and in the quickness with which it melts and its low resistance to fire. The yellow one is more resistant to fire than the red one and its tincture is more abundant, but the red one is the best. Among its properties, it whitens Venus[182] if it is sublimed, and blackens it

[172] I propose this translation on the basis of the meaning of the verb ازق and the context, although this would represent a very rare occurrence. Possible alternative translations are "bottle-necks" and "bellows" (as plural of زق).

[173] The meaning of this expression is ambiguous: the "third transmutation" appears to be highly regarded in the text and could be interpreted as denoting metallic transmutation. Nowhere in the text is any mention made of a progression of transmutations.

[174] The "two leads" probably indicate lead and tin.

[175] The double couples of metals mentioned here could allude to different qualities of the same metal, as in the case of "the two coppers" or different, but similar, metals, like copper and brass or bronze, similarly to the phrase "the two leads" meaning tin and lead.

[176] The name of the final product of this operation is spelled in Syriac letters in the manuscript in a very unclear script: a tentative transcription would be too risky. The context of the passage makes "the Moon," a *Deckname* for silver, the most plausible translation.

[177] I.e. sal ammoniac (see above).

[178] The inferior disk of the head of the distilling apparatus where the sublimed substance would deposit is indicated by the word ترس, literally "lip." See Sébastien Moureau and Nicolas Thomas, "L'aludel: savoir et savoir-faire transmis du monde arabe à l'Occident médiéval?" in *Héritages arabo-islamiques dans l'Europe méditerranéenne*, ed. Catherine Richarté, Roland-Pierre Gayraud and Jean-Michel Poisson (Paris: La Découverte, 2015): 239–52.

[179] The text here describes the two ores of arsenic : realgar, a red ore, and orpiment, a yellow one.

[180] The Arabic text has a *lacuna* here. Ruska, *Das Buch*, 39, suggests supplying the missing text on the basis of the Latin versions, that reads "cum in utrisque non sit fortius eo"; in Arabic, فيهن اغز, "the stronger among them."

[181] I.e. sulphur (see above).

[182] Venus is a common *Deckname* for copper.

fol. 21v[183]

إذا لم يصعد ويحرق ما دام حيا والمراد من العلم تبيضه وإخراج دهانيته وإزالة [a] الإحراق والاحتراق منه وقال جابر في المجردات أنه إذا طبخ بدهن اللوز المر وغيره جاء غاية وزعمت طائفة [b] أن الخل يصلح منه صلاحا لا فساد بعده وأنه عطشان إليه وأنه في الحكمة تمامة وهو لعمري يظهر منه عجائب [c] من الأعمال فافهم ذلك وزعمت طائفة [b] أنه حجر القوم المهان المطروح في المزابل والحمامات لأنه أصل النورة وهو يعقد العبد في الجسد ويثبت به إذا كان معه اخوه وقال في الزيبق السر لا في [184] غيره له [185] وفي المعدنيات أجساد تزاوجه أعيا على كل ذي لب فأعجزه وقيده الحجر المصفر خارجه وتدبيره على ثلث أنواع التصعيد والقتل المقرر حتى يبيض صدق خالد في قوله [§ 11] **صفة تصعيده** وذلك أن تأخذ منه جزأ [d] فتسحقه سحقا ناعما وتأخذ وزنه من ملح الطعام فتسحقه وتحله في مصران ثم روقه وأسقه [e] العلم المسحوق بالسحق والتشوية حتى يحصل في جوفه جميعه ثم اجعله في قدر مزجج وكب عليه آخر واستوثق من الوصل بالصاروج [f] المحكم وطين الشعر وركبه على المستوقد وأوقد عليه النار تحته عشر ساعات ثلاثة [g] بنار لينة

---

a ازالت b طايفة c عجايب d جزوا e اسقيه f بالصاروخ g ثلثة

---

183 A long marginal note, written in a different hand to that of the main text, is found on the right margin of this page. It is devoted to the description of the "fat of the scorpion," a kind of greasy and sulphureous substance.

184 Superscript in the manuscript.

185 Superscript in the manuscript.

fol. 21v

if it has not been sublimed, and it burns as long as it is alive. The aim with the sign is to whiten it, to extract its oiliness, and to remove its flammability. Ǧābir in the [*Book of*] *Abstractions*[186] said that, when it is cooked with bitter almond oil or another oil, it reaches the highest goal. Some people maintain that vinegar improves it in a permanent way, that arsenic is thirsty for vinegar and that it is the perfection of wisdom. I swear upon my life that marvellous operations appear from arsenic: understand it! Some others maintain that arsenic is the stone of the common people, despised and dumped in the pigsty and the toilets since arsenic is the principle of lime,[187] since it makes the servant congeal into a body, and it is fixed with it, if its brother is with it.[188] He said[189] that the secret, that is not found in anything else, is in mercury and that among the minerals there are bodies that marry it. It wears out all the people of reason. The stone whose exterior is yellow weakens it and confines it. Its preparation is of three kinds: sublimation and reiterated killing until it becomes white. Ḫālid was right in his speech.

[§ 11. Sublimation of arsenic]

Description of its sublimation. You should take one part of it, grind it very finely. Then take the same weight of cooking salt, grind it, dissolve it in a piece of gut, and then filter it. Water the ground arsenic with it by grinding and roasting, until it has completely penetrated it. Then put it in a glazed pot and place another pot over it upside down. Seal the joint with reinforced quicklime and clay of the hair.[190] Place the pot on the oven and light a fire under it for ten hours: three on gentle fire

[186] Ruska, *Das Buch*, 31, maintains that the content of the quotation is actually derived from the *Liber Secretum Secretorum* of al-Rāzī, more correctly titled in Arabic *Kitāb al-Asrār* ("The Book of the Secrets"). In the extant Latin versions the book's title is given as *Liber de spoliationibus* (Steele's edition) and *Liber denudatorum* (Ruska's edition). On the alchemical books and authorities mentioned in the Arabic *OAS*, see *Introduction*, § 4.3.

[187] It is not clear what the expression "principle of lime" is intended to denote here. The Latin versions have *radix calcis* (Ruska ed. § 1) that corresponds to the Arabic reading, and *radix noce* (Steele ed. § 11). In the Hebrew translation we find the expression "root of lime." The idea that the prime matter of the stone is commonly despised and deprecated by ignorant people who do not recognize its value is a common trope in alchemical literature.

[188] The idea that mercury does not die unless "with its brother" is a common trope in Latin medieval alchemy.

[189] This quotation, that a reader might attribute to Ǧābir as in the previous section, is actually from Ḫālid ibn Yazīd, who is praised as truthful at the end of the passage, and belongs to line six of poem 91, according to the classification proposed by Marion Dapsens, "The Alchemical Works of Khālid," 368. I am very grateful to Marion Dapsens for identifying this quotation. On the alchemical books and authorities mentioned in the Arabic *OAS*, see *Introduction*, § 4.3.

[190] This insulating material composed mainly of fermented clay and hair is commonly known as "the clay of wisdom" (*lutum sapientiae* in Latin). On the composition of the *lutum sapientiae*, see Nicolas Thomas, "De la recette à la pratique, l'exemple du *lutum sapientiae* des alchimistes," in *Craft Treatises and Handbooks: The Dissemination of Technical Knowledge in the Middle Ages*, ed. Ricardo Cordoba de la Llave (Turnhout: Brepols, 2013), 249–70.

fol. 22r

وسبعة بنار قوية وإذا برد فاجمع ما صعد بصوفة فإن لم يرضيك بياضه فاعمل عليه التدبير حتى يرضيك وذلك في ثلث مرار فأبدل له الملح في كل مرة يخرج مثل الكافور والسر في معرفة إدراك المصعدات أن تنثر منه على صفيحة ☾ محمية لا يسودها [§ 12] **صفة غسله** أن تستقيه بالملح المحلول كما تقدم وتشويه في فرن خبز ساعة بعد ساعة حتى يستوعب وزنه من الملح المحلول ثم اغسله عن الملح واجعله في قدر مزجج كلما يخرج الماء مالحا صفه [a] حتى يخرج حلوا [b] ليس فيه ملوحة فجففه وشمعه بملح القلي المحلول ويكون وزنه حتى تراه أبيض مثل الثلج شحم فحله في نار لينة حتى ينحل ويجمد فارفعه [§ 13] **آخر** أجل منه خذ من العلم الأصفر المصفى أعنى الصفائح [c] جزا [d] ومن ملح العجين مثله فحله في مثله من خل حادق وأسق العلم منه بالسحق والتجفيف شيئا بعد شيء حتى يستوعبه واشوه [e] واغسله بالماء العذب[f] حتى يذهب عنه الملوحة وتراه أبيض مثل الثلج ثم أحرق به أي جسد شئت فإنه غاية [§ 14] **صفة تدبيره** بالتقرير أن تأخذ منه جزا [d] فتسحقه ناعما [g] في هاون وخذ من ملح الطعام بعد تشويته وأذبه واخدمه بالسقي والتجفيف في الشمس الحارة

---

[a] صفيه [b] حلو [c] صفايح [d] جزو [e] اشويه [f] العدب [g] ناعم

fol. 22r

and seven on strong fire. When it cools, collect what sublimed with some wool. If you are not satisfied with its whiteness, perform the operation on it until you are satisfied. [Do] this three times, replacing the salt each time; it comes out like camphor.[191] The secret in knowing how to obtain sublimates is that you scatter some of it on a warm sheet of ☾,[192] that will not blacken.

[§ 12. Improvement of arsenic (i)]
Description of its washing. You water it with dissolved salt, as mentioned, and roast it in a bread oven hour by hour until it has absorbed the equivalent of its weight of dissolved salt. Then wash away the salt from it and place it in a glazed pot. Every time the water comes out salty, purify it until it comes out sweet with no saltiness in it. Then dry it and cerate it with the same weight of dissolved alkali salt until you see it white as snow. Grease it and then dissolve it over a gentle fire until it is dissolved and congealed. Take it up.

[§ 13. Improvement of arsenic (ii)]
Another [operation], better than the previous. Take one part of purified yellow sign, namely flakes,[193] and as much dough salt. Then dissolve it in the same amount of sharp vinegar. Water the sign with it by grinding and drying[194] it little by little until it has absorbed it. Roast it and wash it with fresh water until the saltiness has left it and you see it as white as snow. Then burn any body you want using it. This is the [highest] goal.

[§ 14. Making silver with arsenic]
Description of its preparation through fixation
You should take one part of it and grind it finely in a mortar. Take some kitchen salt after roasting it, dissolve it and work it by watering it and drying it in the hot sun

[191] The sentence does not appear to describe a possible behaviour of arsenic. In the translation, I assume that a conjunction ف between the two phrases may have been lost in the transmission of the text. A literal reading of the passage as it stands would be "replacing the salt each time it comes out like camphor."

[192] The manuscript here uses the symbol of a crescent to indicate the Moon, i.e. silver.

[193] The Arabic text here prescribes the use of *ṣafā'iḥ* of arsenic, that can be translated also as "sheets," "plates," or "leaves." I have translated it as "flakes" to reflect the physical appearance of mineral orpiment.

[194] This expression recurs several times in the text: the Latin versions have "cum contritione et exsiccatione" (Steele ed. § 14) and "cum molitione et siccatione" (Ruska ed. § 4); the Hebrew version reads "by crushing and drying" (§ 15).

fol. 22v[195]

حتى يستوعبه فاعجنه بوزنه من الدهن المذكور وضع الكل في زجاجة وطينها واجعلها في فرن الخبز وألق عليها الغطاء [a] ليلة واحدة واسحقه وكرر ذلك حتى تراه جسدا أبيضا [b] مثل القمر فاصبغ به ما شئت [c] تصل إلى مرغوبك بحول الله وقوته وألق منه درهما [d] على علم على من زهرة[e] أو مريخ أو أسرب تلحفها ويلقي عليها بنار المكين والفحم حتى يذوب قمرا طيبا [f] فاعمل كيف شئت[c] [§ 15] **القول في العقرب** وطبعه كطبع العلم وتدبيره كتدبيره على ما ذكره أهل الصناعة في كتبهم ولم أجربه ولكني جربته مع العبد وصعدته وأتاني زنجفرا عجيبا [g] وجربته في إحراق الزهرة فأتاني راسختا عجيبا على ما أذكره إن شاء الله والمراد منه إزالة [h] إحراقه وذهاب دهانيته وتبيضه مثل العلم سواء[i] فأحسنه الأحمر وذكر جابر في الرياض أن العلم يقوم مقام العقرب في الحمرة والعقرب يقوم مقام العلم في البياض ولم أجربه [§ 16] **صفة تدبيره** فاسحقه بوزنه شبا يمانيا وألق عليه من البول المعتق ما يغمره واطبخه بنار لينة واغله [j] فيه حتى يذهب الماء ويرضيك لونه وبياضه وإلا فعد عليه من ماء الجبن الحامض شيئا وكرر ذلك عليه حتى يبيض وأدخله فيما تريد أبشر به

---

[a] الغطا [b] جسد ابيض [c] شيت [d] درهم [e] زهره [f] قمرطيب [g] زنجفرعجيب [h] ازالت [i] سوي [j] اغليه

---

[195] A long marginal note, written in a different hand from that of the main text, appears on the right margin of this page. It describes the preparation of a sharp water used for breaking iron.

fol. 22v

until it has absorbed it. Then knead it with the same weight of the aforementioned oil and put everything in a glass flask. Coat it with clay, place it in a bread oven, and put the lid on it for one single night. Grind it and repeat this until you see it as a body white like the Moon.[196] Dye what you want with it: you will reach the object of your desires, with the help of God and His power. Throw one *dirham* of it on the sign [and then] on Venus, Mars, or lead, covering them and applying a strong fire of charcoal until it melts into good Moon, and then do as you like.

[§ 15. On sulphur]
Discourse on the scorpion. Its nature is like the nature of the sign and its preparation is like the preparation [of arsenic] according to what the people of the Work[197] mentioned in their books, but I have not tried it. However, I have tried it instead with the servant and I have sublimed it: it gave me wonderful cinnabar. I tried it also for the combustion of Venus and it gave me wonderful *rāsḫat*,[198] as already mentioned, if God wants. The aim with sulphur is to stop it from burning, to remove its oiliness and to whiten it, just as if it were the sign. The best kind is the red one. Ǧābir in the *Flower Beds*[199] said: "The sign takes the place of the scorpion for the red, while the scorpion takes the place of the sign for the white." But I have not tried it.

[§ 16. Preparation of sulphur]
Description of its preparation. So, grind it in with the same weight of Yemenite alum,[200] throw enough matured urine on it to cover it and cook it on a gentle fire. Boil it in this until the water goes away and you are satisfied with its colour and its whiteness. If not, then place on it some sour whey and repeat this [process] on it until it is whitened. Introduce it in whatever [preparation] you want. Rejoice with it,

[196] The Moon is a common *Deckname* for silver.
[197] People of the work, *Ahl al-ṣināʿa*, refers to the alchemists.
[198] *Rāsḫat* designates burnt copper, the Latin *aes ustum*.
[199] The *Book of the Flower Beds* (*Kitāb al-Riyāḍ*) is the genuine title of a Ǧabirian work. On the books and authorities mentioned in the Arabic *OAS*, see *Introduction*, 4.3.
[200] Alum from Yemen, white and flaky, was praised for its purity and is a very common ingredient in alchemical as well as medical recipes in the medieval Arabo-Islamic world: see Moureau, *De Anima*, 252–3. Frequent mentions of Yemenite alum are found, for instance, among the medical fragments from the Cairo Genizah, where alum features in lists of goods imported from Yemen.

fol. 23r

إن شاء الله تعالى [§ 17] **صفة** [201] ✎ خذ من الزاج الطيب بعد تشويته في الفرن وانقعه في مثله من بول المعتق واتركه فيه حتى ينحل الكل فصفه[a] حسنا واجعله في زجاجة واجعل فيها مثل وزن الزاج من برادة المريخ واتركه ٧ أيام حتى تنحل البرادة كلها فخذ من الزهرة المحرقة ما شئت واسحقها بها وأسقها [b] منه شيئا بعد شيء ودم عليه السحق حتى يستوعبه كله فألق [c] منه واحد على أربعة من القلعي يخرج ✎ [202] خالص يأتيك غاية بإذن الله [§ 18] **صفة** تأخذ راسخت ٥ توتيا ٥ دم اخوين قاطع واحد زعفران نصف كركم واحد زاج [d] قبرصي واحد زنجفر [e] جزأ [f] رمان حلو جزأ [e] شعر جزأ [e] ويسحق ويكون بطال [§ 19] **صفة** ☾ تأخذ وزن درهم ☾ برادة وتلغمها بعشرة عبد وتلغمها بمقرعة حديد في خل بكر وتحركه في بسمار على نار حامية وكل ما نقص الخل تزيده حتى يعقد ويبقى [g] معقودا خذ من ذلك العبد تزينه ومهما كان وتحط لكل وزن عشرة ؟ علم أحمر وتلغمه في بعضه بعضه وتصره في خرقة من شاش يابس [h] وتر بطها بخيط قنب وتدير شحم خنزير صافي وتسقطه فيه الزيوغ با لمقرعة [203] ويكون الشحم غامر الطيار به المربوط ولا يكون تخلي يصل إلى أسفل وأنت ماسك الخيط والنار عمالة والعبد يصيح حتى ينقطع صياحة علامة

---

[a] فصفيه [b] اسقيها [c] فالقى [d] جاز [e] جنزفر [f] جزو [g] يبقا [h] يانس

---

[201] An oblong symbol (which only slightly resembles the font I have employed above) is here used to indicate gold.

[202] The same symbol is here employed to represent the result of this operation.

[203] The preposition and the first element of the article that should be attached to this noun are detached from it and conclude the previous line of text.

fol. 23r

if God, may He be exalted, wants.

[§ 17. Making gold with vitriol, iron and copper]
Description of ✎. Take good vitriol, after roasting it in the oven, and soak it in the same amount of matured urine. Leave it in the urine until all of it has dissolved. Purify it well, place it in a glass flask, and put in it the same weight as the vitriol of filings of Mars. Leave it there for seven days until all the filings are dissolved. Then take as much as you want of burnt Venus, grind it in this and water it little by little. Keep on grinding it until it has absorbed everything. Throw one [part] of this on four [parts] of *qalaʿī*: it will come out pure ✎,[204] that will give you the [highest] aim, with the permission of God.

[§ 18. Making gold with burnt copper]
Description. Take five [parts of] *rāsḫat*, five [parts of] tutty, one [part of] sharp dragon's blood,[205] half [a part of] saffron, one [part of] turmeric, one [part of] Cypriot vitriol, one part of cinnabar, one part of sweet pomegranate, and one part of hair. Grind them and it will be useless.[206]

[§ 19. Making silver with mercury, vinegar and arsenic]
Description of ☾.[207] Take the weight of one *dirham* of filings of ☾, amalgamate them with ten [*dirham*] of servant, mixing them with an iron stick in fresh vinegar. Stir it in a nail[208] over a hot fire and every time the vinegar diminishes, add it [again] until it coagulates and stays like that. Take some of that servant and embellish it. Whatever happens, take for every ten [*dirham*] in weight, five[209] [*dirham*] of red sign, amalgamate one with the other and wrap around it a cloth of dry muslin. Fasten the cloth with a hemp thread, spread pure pork fat around it, and put the *zayyuġ* into [the enclosed material] using a stick.[210] The fat will hold the volatility bound to [the enclosed material] and it will not fail to reach the centre. While you hold the thread and the fire is active, the servant screams until its scream breaks off, a sign

[204] The same symbol employed above is repeated here to represent the product of this operation.

[205] Dragon's blood is the product of the dragon's blood tree. See Max Meyerhof, *Sharḥ asmāʾ al-ʿuqqār. L'explication des noms de drogues* (Cairo: Imprimerie de l'Institut Français d'archéologie orientale, 1940), n° 96.

[206] The Arabic text here employs the word *baṭṭāl* ("idle," "inactive," "unemployed"), which is not found elsewhere in the text.

[207] The symbol of the crescent, which is often found in the manuscript of the *OAS* generally represents silver and it is interpreted as such in my translation.

[208] *Bismār* in the manuscript, a colloquial form for *mismār* ("nail"). The context would indicate that, rather than a nail, a vessel or pot of some kind should be employed here.

[209] The figure in the manuscript can be read as the Eastern form of the Arabic numeral 5. On the different forms of Arabic numerals, see Rida A. K. Irani, "Arabic Numeral Forms," *Centaurus* 4 (1955): 1–12; Charles Burnett, "Indian Numerals in the Mediterranean Basin in the Twelfth Century, with Special Reference to the 'Eastern Forms,'" in *From China to Paris: 2000 Years' Transmission of Mathematical Ideas*, ed. Yvonne Dold-Samplonius, Joseph W. Dauben, Menso Folkerts and Benno Van Dalen (Stuttgart: Steiner 2002), 237–88; reprinted in Charles Burnett, *Numerals and Arithmetic in the Middle Ages* (London-New York: Routledge, 2010) and Richard Lemay, "Arabic Numerals," in *Dictionary of the Middle Ages*, ed. Joseph Strayer (New York: Charles Scribner's Sons, 1982), 382–98.

[210] The text here is problematic and my translation, therefore, provisional. The passage appears to describe the process of smearing and pressing fat onto the cloth containing the mixture of silver filings, mercury, and arsenic in order to prevent the volatile component (the mercury, or "servant") from escaping. The word *zayyuġ* could be translated as "deviations," which does not seem to be fitting in the context, or as a rather unusual plural of *zāġ* ("crow") and therefore as a *Deckname*.

fol. 23v

بموته فإنه يخرج ☾ [§ 20] **في حل الزجاج** [211] من غير تعب يؤخذ الزجاج و يسحق ناعما [a] هباء ويضاف نشادر ونطرون وعقاب أجزأ سواء [b] تسحق على نار لينة بمقرعة [c] فإنه يصير ماء لا يجمد أبدا ويعقد العبد ثابتا [d] وازوج فيه القلعي إقامة تم [§ 21] **صفة القلعي** يؤخذ آنك وطفى في جير ٧ مرات [e] وفي عسل ٧ وفي قطران ٧ وفي زيت ٧ وفي خرئ بقر الأخضر ٧ وبعد ذلك شب يماني وبياض بيض وقليل زيت لته فيه واطفه [f] سبع مرات [g] يقوم بعون الله [§ 22] **صفة تدبير** في الطلق يحمي الطلق ويغمس في ماء الفجل ٣ مرات [g] ويدق فإنه يندق كالكحل ثم يجعل في قدح ويصب عليه ماء الفجل غمرة وأربع أصابع ويوضع في الشمس عشرة أيام ينحل مثل الزيبق ثم يؤخذ مثله عبد ومثل الجميع مرقشيثا ويؤخذ بوزن الأدوية تنكار ويسحق بماء الفجل حتى يختلط ويصير شيء واحد ثم يسبك فيصير حجر أبيض يلقي منه درهم على ? رصاص وعلى الزهرة يصير ☾ إن شاء الله تعالى [§ 23] **باب عمل سكر** يقطر الدبس ثم يغلي منه ١٠ أواق إلى أن يبقى [h] مقدار ستة أواق ويأخذ منه جزأ [i] ومن النشا [در] [212] جزأ [j] ومن السكر جزأ [k] ويعقدهم يجي غاية وهو مجرب [§ 24] باب **عمل المسك** يؤخذ دم تيس أسود ليس يكون فيه شيئا أبيض وتتركه في الشمس

---

[a] ناعم [b] سوي [c] في مقرعة [d] ثابت [e] مراة [f] اطفيه [g] مراة [h] يبقا [i] جزو [j] جزوا [k] جزو

---

[211] The copyist here appears to be confused about the title and topic of this section: he first wrote زاج (vitriol), but later corrected it to زجاج (glass) and repeated this word in separated Arabic letters on the upper margin. This section is absent from the Hebrew version of the *OAS*.

[212] The last two letters of the word نشادر are not present in the manuscript.

fol. 23v

of its death. And so ☾[213] will come out.

[§ 20. Melting glass]

On the melting of glass with no effort.[214] You should take the glass, grind it finely to a powder, and add sal ammoniac, natron, and eagle in equal parts. Grind [it] over gentle fire with a stick and it will become water which will never congeal, which coagulates the servant [making it] fixed, and with which the *qalaʿī* will unite through rectification. The end.

[§ 21. Rectification of *qalaʿī*]

Description of the *qalaʿī*.[215] You should take tin and temper it[216] seven times in lime, seven times in honey, seven times in tar,[217] seven times in oil and seven times in green cow manure, and then in Yemeni alum and egg white and a little oil. Mix it with this and temper it seven times: it will be rectified with the help of God.

[§ 22. Making silver with talc, mercury, borax]

Description of the operation on talc. One should warm up talc, dip it three times in water of radish,[218] and pulverise it, so that it will be pulverised like *kohl*.[219] Then one should put it in a bowl and pour on it enough water of radish to cover it by four inches . It is placed in the sun for ten days: it will dissolve like mercury. Then one should take the same amount of servant, the same amount of marcasite as the whole, and the same weight of borax as the [other] ingredients,[220] and one should grind these with water of radish until they have mixed and become a single thing. This will be melted and so it will become a white stone. One should throw one *dirham* of it on five[221] [*dirham*] of lead[222] and on Venus. It will become ☾, [223] if God, may He be exalted, wills.

[§ 23. Preparation of sugar]

Chapter on the work on sugar. One should distil the syrup[224] and then boil ten ounces of it until the amount of six ounces remains. One should take one part of it, one part of sal ammoniac, and one part of sugar, and congeal them. It will produce the [highest] goal. This is tested.

[§ 24. Preparation of artificial musk]

Chapter on the work on musk

One should take the blood of a black billy goat in which there is nothing white. Leave it in the sun.

[213] The symbol of a crescent here represents the final product of the previous procedure, namely silver.

[214] I here translate the verb حل , which I normally render as "to dissolve," with the verb "to melt," to reflect the kind of operation described and the behaviour of glass.

[215] In Arabic alchemical, medical, and mineralogical literature, the word *qalaʿī* is used to define different kinds of tin and lead. Its meaning is not always consistent as it can describe pure tin, pure lead, black tin, and black lead, but also white tin and white lead. Given this ambiguity, I have preferred to leave the term untranslated. See Reinhart P. A. Dozy, *Supplément aux dictionnaires arabes* (Leiden – Paris: Brill – Maisonneuve 1927), vol. II, 397, where the word is etymologically connected to the name of an Indian city, *Qalaʿa* or *Kala*, where tin was mined. See also Fabian Käs, *Die Mineralien in der arabischen Pharmakognosie. Eine Konkordanz zu mineralischen Materia medica der klassischen arabischen Heilmittelkunde nebst überlieferungsgeschichtlichen Studien* (Wiesbaden: Akademie der Wissenschaften und Literatur, 2010), on 201–4, 223–26, 293–96, 582–86, 901–3. For a discussion and a tentative assessment, see Moureau, *De Anima*, I, 273.

[216] The procedure described in this passage is not very clear: the sequence of substances could function as a superficial mordant or polish for tin.

[217] قطران , generally translated as "tar," can also designate the sap of the juniper trees that was cooked and used to smear the mantle of camels; other possible meanings are "melted brass" and "any melted substance."

[218] This kind of water only appears in this passage of the *OAS* and is not a common ingredient in Arabic alchemical treatises. It is not clear whether this water is obtained by infusion or by distillation or if the whole should be interpreted as a *Deckname*.

[219] Kohl (كحل) is a preparation of pulverised stibnite (antimony) used as a cosmetic and medicine for the eyes since ancient Egyptian times. Its use is recorded all over the Near East, the Mediterranean basin, and parts of Africa.

[220] *Adawiyya* in the text, lit. "remedies," "medicines."

[221] The character in the manuscript can be read as the Eastern form of the Arabic numeral 5. See n. 212, above.

[222] I translate رصاص as "lead" throughout the text. It should, however, be noted that this word is also often used for tin. See § 8 below, where "the two leads" are mentioned; these "two leads" possibly indicate both lead and tin.

[223] I.e. silver.

[224] The word دبس indicates a dense sugary paste, like syrup, treacle, or molasses.

fol. 24r

[§ 25] سبعة [225] ٧ أيام فإنه ينحل ماء أحمر مثل الدم فشمع به العبد الذي صعدته أولا عن الزاج والملح وأدم عليه السحق والسقي والتجفيف حتى تستوعبه ثم أدخله للحل [a] ثم أعقده بعد ذلك اطرح منه درهم على حا [226] من القمر يخرج ⬮ [227]

[§ 26] **القول في القمر** اعلم أن القمر جسد نقي لكنه أنقص من الشمس وهي من قسمة القمر وضياه دون ضياء الشمس ورتبته كذلك وليس له كما لها في ضيائها وقوتها وكذلك منزلة القمر عند الشمس لكن القمر يفسد في الأرض وفي النداء وطبعه إلى الحموضة فلذلك يتزنجر كالنحاس وتلحقها الآفات وتحرقها الكباريت وتنقص في النار وليس بعد الشمس جسد أكرم منها وهي أقرب الأجساد إلى الشمس لأن باطن الشمس ظاهر القمر وظاهر القمر باطن الشمس وطبعها باردة رطبة وقيل باردة ويابسة وتخلط مع الشمس والزهرة وتقبل الصبغ وقد وصفنا منها إكسير البياض والحمرة وميزناهما [228] والآفات اليها سريعة لأنها رفيعة وليس لها صبر على ما يؤذيها والكباريت لها عدوة تؤذيها ولا توافقها والكباريت نفورة في تدبير الزهرة والقمر والأܣܪܒ [229] والعلم للذهب والمريخ والآنك فاعلم و القمر جسد الأبيض في العمل الأعظم وقد يكون يقوم مقامها القلعي في العمل الأصغر والعلم نفسها والعبد روحها والعقاب خادمها لها يؤلف بينهم

---

[a] الحل

---

[225] The number seven is given both in letters and as a numeral.

[226] In the manuscript, a separate letter ح followed by a ا is found here; this feature makes it difficult to decipher this passage.

[227] The manuscript here has an oblong shape that should stand for the name of the product of the operation previously described.

[228] This reading is conjectural and based on the context. The Arabic text preserves Syriac letters, a *heh*, an *aleph*, and possibly a final *heh* or *semkhat*.

[229] The name of this ingredient is spelled in Syriac letters in the manuscript. It could be read as *usrub* ("lead").

fol. 24r

[§ 25. Preparation of gold (vi)] [...][230] for seven [7][231] days: it will dissolve into a water red like blood. Then cerate with it the servant that you previously sublimed from vitriol and salt, and continue the grinding, watering, and drying until it has absorbed it. Then make it dissolve. Afterwards, congeal it and after this throw one *dirham* of it on ‡ of Moon. ‡ will come out.[232]

[§ 26. On silver]
Discourse on the Moon. Know that the Moon is a pure body, but it is less [pure] than the Sun. It is from the part of the Moon and its luminosity is inferior to the luminosity of the Sun, as is its degree, also.[233] As for its luminosity and faculty, the Moon does not have it as the Sun does: such is the rank of the Moon in relation to the Sun. But the Moon corrupts itself in the earth and in humid places and its nature [tends] towards sourness. For this reason, it is corroded like copper, blemishes cover it, the sulphurs burn it, and it diminishes in the fire. After the Sun, there is no body nobler than it. It is the closest body to the Sun, since the interior of the Sun is the exterior of the Moon and the exterior of the Moon is the interior of the Sun. Its nature is coldness and humidity.[234] It is [also] said that it is coldness and dryness. It mixes with the Sun and Venus and receives the tincture. We already described [how to obtain] from it the elixir of whiteness and of redness and we have distinguished them. Blemishes are quick to [attack] it because the Moon is subtle and does not endure what damages it. The sulphurs are for the Moon an enemy that damages it and they are not suitable for it. The sulphurs are terrible in the operation of Venus, of the Moon, of lead,[235] of the sign for gold and of Mars and tin. Know it! The Moon is the body of the white in the great work; the *qalaʿī* may take its place in the lesser work. The sign is its soul, the servant is its spirit, and the eagle is its attendant that unites them

[230] The text of this recipe does not represent the conclusion of the chapter on the musk, but is actually the end of the closing recipe of the section on gold that is not preserved in the Arabic manuscript, but is available in the Latin and Hebrew translations of the *OAS*.

[231] The number seven is given both in letters and as a numeral in the manuscript.

[232] The reading of this passage is complicated by the presence in the manuscript of symbols and abbreviations that prevent the understanding of the ingredients and products of this procedure.

[233] The passage plays with the real and metaphorical meaning of Sun and Moon, here standing for the metals gold and silver in the first occurrence, and for the actual planets in the second.

[234] In the discussion of the internal and external natures of gold and silver, the author attributes a masculine nature to the sun, which is grammatically feminine in the Arabic language, and a feminine nature to the Moon, which is masculine. A further element of shift in genders is the attribution adjectives of feminine form to the masculine word *ṭabʿ*.

[235] A word in Syriac letters is found here in the manuscript. It appears that it could be read as a transliteration from the Arabic اسرب , meaning "lead," but lead is mentioned in this same sequence, suggesting another interpretation.

fol. 24v

ويثبته [§ 27] **صفة** تدبيرها أذبها في بوط وألق [a] عليها وزنها قلعي ثم صبه في العجل واسحقه وارفعه [§ 28] **آخر** خذ الأسرب القلعي أذبه في بوط وصبه في ماء وملح مرارا ثم أحم [b] القمر وصبه في ذلك الماء بعد أن تجعل في كل رطل من الماء درهم من العقرب أو من العلم فإنه تتكلس سوداء فعالجها بالملح المحلول واشويها في الأتون ثم اغسلها من الملح فأنها تبيض وتتهبأ [§ 29][c] **آخر** وأن طبخ العقرب بالماء والملح ثم سقيت به برادة الفضة[236] وأدخلها في الأتون[237] ليلة ثم اغسلها عن الملح صارت كلسا أبيضا[d] في ثلث مرات [§ 30] **آخر** خذ برادة ما شئت فلتها بالزحل ثم أوضعها على شقفة جمر فإذا وصل إليها اللهيب ارم عليها شيئا من العقرب حتى تحترق وتصير رمادا فاسحقها بالملح المحلول على ما تقدم وارفعه ذرورا أبيض [§ 31] **آخر** خذ طرق الفضة[238] صفائح [e] أرق ما تقدر عليه وأحمها [f] وأطفها [g] في زحل قد أذبت فيه شيئا [h] من العقرب وكرر عليه العمل حتى تراها [i] سودا فاسحقها بالملح المحلول كما تقدم وارفعها ثم خذ من غبارها جزأ [j] ومثله علم فشمعها بالشب المحلول في البول مع مثله عقاب بالسحق والسقي حتى يتم الماء في جوفها ثم اجعل الكل في بيضة دجاجة فارغة مطينة بطين الحكمة واجعل على فمها قشرة أخرى وطين بالطين المحكم

---

[a] القى [b] احمي [c] تتهيا [d] ابيض [e] صفايح [f] احميها [g] اطفيها [h] شيء [i] تريها [j] جزو

---

[236] The letter ض in this word is here replaced with a Syriac letter *teth*.
[237] The final ن in this word is stretched and its trait reaches to the following four words.
[238] The letter ض in this word is here replaced with a Syriac letter *teth*.

fol. 24v

and makes them stable.

[§ 27. Preparation of silver (i)]
Description of its operation. Melt it in a crucible and throw on it the same weight of *qalaʿī*. Then pour it quickly, grind it and take it up.

[§ 28. Preparation of silver (ii)]
Another [operation]. Take the *qalaʿī* lead, melt it in a crucible, and pour it repeatedly into water and salt. Then heat the Moon and pour it into that water after adding for each *raṭl* of the water a *dirham* of the scorpion or of the sign. It will then become a black calx. Treat it with dissolved salt, roast it in the oven, and then wash it from the salt. It will whiten and become a powder.

[§ 29. Preparation of silver (iii)]
Another [operation]. You should cook the scorpion in water and salt and then imbibe with it the filings of silver. Place it in the oven for a night and then wash it from the salt. It will become a white calx after three times.

[§ 30. Preparation of silver (iv)]
Another [operation]. Take as much as you want of [silver] filings and then mix them with Saturn.[239] Then place it on a pot sherd in embers. When the flame reaches it, throw on it some scorpion until it burns and becomes ash. Then grind it with dissolved salt according to what is [explained] above and take it up: [it will be] a white powder.

[§ 31. Preparation of silver (v)]
Another [operation]. Take silver and hammer it in sheets as thin as possible, warm them, and temper them in Saturn in which you have already melted some scorpion. Repeat the work on it until you see it is black. Grind it with dissolved salt as you did previously and take it up. Then take a part of their powder and as much of the sign. Cerate them with alum dissolved in the urine [and] with as much eagle, grinding and watering, until the water in them is finished. Then put everything in an emptied chicken egg coated with clay of wisdom. Put on its opening another eggshell, coat it with reinforced clay,

[239] Saturn is a common *Deckname* for lead.

fol. 25r

وجففها واجعلها في نار زبل البقر نصف نار لا تكون جحيما واتركها فيه يوم أو ليلة ثم أخرجها تجدها نقرة معقودة فألق [a] منها درهم على ٧ من ١ ٣٠ ملح [240] تخرج ☾ إن شاء الله تعالى [§ 32] **آخر** أجل منه خذ من صفائح [b] ٣٠١ قمر [241] جزأ [b] الرقاق ومن العلم المدبر جزأ [c] ومن الأجساد أربعة فالحفها به وأودعها في بيضة دجاجة مطينة بطين الحكمة ودعها في قدر برام واجعلها في فرن الخبز يوم وليلة ثم أخرجه وشمعه وشمع ما فيه بماء الشب والعقاب حتى تراه كالشحم فألق [d] منه على القلعي يقوم وإن أخذت القمر غبار مسحوق مع العلم وسحقته وأدخلته الإحراق كان عملك أجل وأرفع فاعمل بأيهما شئت تصل اليه إن شاء الله [§ 33] **القول في القلعي** وهو الأسرب الصرفان والأسرب القلعي وهو من قسمة المشتري وطبعه الحرارة والرطوبة لكنه قليل الحرارة لأنه مزاجه انفسد بغلبة العقرب عليه في معدنه وفيه الرخاوة والصرير وسرعة الذوب والحموضة فمن أزال عنه هذه العلل الأربع رجع قمرا [e] بإذن الله وهو يقبل الصبغ ويثبت فيه ويبيض الزهرة فضة لأنه منفرد بها ويقبل صبغ الحمرة ويكون منه ⬯ [242] مرتفعة ويصنع منها ماء شريف فيحبس به العبد وهو عدو القمر والزهرة ملائم [f] الشمس والمريخ [§ 34] **صفة تدبيره**

---

[a] فالقى [b] صفايح [c] جزو [d] فالقى [e] قمر [f] ملايم

---

[240] The word for salt appears written in Syriac letters in the manuscript, preceded by the numerals 1 and 30. The two figures can be decoded, according to the system of correspondences of numbers and Arabic letters, as the article *al*. On the left margin of the page, a note in separate letters appears to erroneously interpret the Syriac letters for "salt" as the word as *qala'ī*.

[241] The word for Moon is written in Syriac letters in the text, preceded by the numerals 1 and 30. As in the previous line, these two figures can be decoded as the article *al*. On the left margin of the page, a note in separate letters appears to correctly decode the Syriac word as قمر, "the Moon," i.e. silver.

[242] The text here preserves an oblong symbol representing the product of the previous operation.

fol. 25r

and dry it. Put it in fire of cow manure, half fire that is not scorching, leave it in it for a day and a night, and then take it out. You will find it like a congealed cavity.[243] Then throw one *dirham* of it over seven [*dirham*] of the salt: ☾ [244] will come out, if God, may He be exalted, wills.

[§ 32. Preparation of silver (vi)]
Another [operation] better than this. Take one part of fine sheets of the Moon, one part of the prepared sign, and four [parts] of the bodies, and cover them with it. Place it in a chicken egg coated with clay of wisdom, place the egg in a saucepan, and put it one day and one night in a bread oven. Then take it out, cerate it, and cerate what it contains with water of alum and of the eagle until you see it like fat. Then throw [some of] it on the *qalaʿī* and it will be rectified. But if you take the Moon [in] powder crushed with the sign, you pulverise it and introduce it in the combustion, then your work is better. Take it up and work with any of the two you wish: you will reach [your goal], if God wills.

[§ 33. On *qalaʿī*][245]
Discourse on the *qalaʿī*. It is the lead *ṣarafān* and the lead *qalaʿī*. It is from the part of Jupiter and its nature is hotness and humidity, but it has a limited humidity, since its composition was damaged by the predominance of the scorpion over it in its mineral.[246] In it are softness, squeaking, quickness in melting and acidity. He who removes from it these four illnesses, recovers Moon, with the help of God, while it receives the tincture that becomes stable in it. It whitens Venus [into] silver, because it is separated from it.[247] It receives the tincture of the red and from it comes superior ⸜ .[248] A noble water is produced from it, with which the slave is imprisoned. It is an enemy of the Moon and Venus, [but] suitable to the Sun and Mars.

[§ 34. Making silver from *qalaʿī* (i)]
Description of its operation.

[243] This passage possibly refers to the shape the ingredients previously inserted in the eggshell will assume after prolonged cooking: they will congeal on the interior walls of the eggshell leaving an empty space in the middle.

[244] The crescent represents the Moon and, therefore, silver as the product of this operation.

[245] This term is often translated as tin, but in this text it can indicate different metals.

[246] Tin's nature is here explained with reference to the mercury-sulphur theory: during the formation of tin's mineral below the ground, the overabundance of sulphur, which is naturally hot and dry, reduces the proportion of tin's moist component. On the mercury-sulphur theory, see above § 3.2.

[247] The Arabic text is ambiguous, since the expression could also be rendered as "it is unparalleled with it." The Hebrew version (§ 55) has "because it is similar to it." The Latin translations render this passage as: "quia est singulari ei" (Ruska ed. § 55) and "quoniam est cum [sic] singulare cum eo" (Steele ed. § 65).

[248] The text here preserves an oblong symbol representing the product of the previous operation.

fol. 25v[249]

خذ منه جزأ [a] فاذبه في مغرفة حديد وألق [b] عليه ملحا مسحوقا وحركه بحديدة وحتى يصير غبارا ثم أخرجه واسحقه شحقا ناعما ثم اجعله في إناء فخار و سد رأسها واعمل عليه الطين المحكم وبيته في فرن الخبز ليلة وأخرجه وأغله [c] بما يغمره من الماء العذب [d] واتركه حتى يرسب ثم أهرق عنه الماء المالح وأعد عليه العمل بالماء العذب [d] وحركه تحريكا حسنا واتركه أيضا حتى يرسب وأهرق عنه الماء افعل به ذلك مرارا حتى لا تجد فيه من الملح طعما ولا ترى [e] فيه سوادا ثم نشفه حتى يجف ويصير كلسا ثم خذ من الشب اليماني مثله فاسحقه وانقعه في مثله خلا حادقا وبولا معتقا وألق [f] عليه وزنه عقابا مصريا واتركه فيه حتى ينحل فشمع الكلس بهذا الماء شيئا [g] بعد شيء وأنت تسحقه وتسقيه وتجففه في الشمس يكون سقيك له نقطة بعد نقطة ولا تسقيه دفعة حتى يتشمع ويترطب وتراه المرة العاشرة او الحادية عشر قد بلغ من اللين واللزوجة إلى أن يلتصق به الفهر في الصلاية من لينه وتلزجه فخذ من الزهرة أوقية وأذبها في بوط وألق [h] عليها درهما [i] وانفخ عليها حتى يذوب ويغوص فيها الدواء فإنه يخرج قمرا بإذن الله [§ 35] **وجه آخر** يؤخذ عشرة دراهم قلعي يذاب [j] في مغرفة حديد ويلقي عليها درهم قمر [250] بدرهم حجر [251]

---

[a] جزو [b] القى [c] اغليه [d] العدب [e] ترا [f] القى [g] شيء [h] القى [i] درهم [j] يداب

---

[249] Two lines of marginalia are found on the right margin of the manuscript, corresponding to lines 6–8: this note lists ingredients (Mars, scorpion, eagle) and says that they enter into a better preparation, probably in comparison with the one in the main text.

[250] In Syriac letters in the manuscript. The corresponding Arabic word قمر (Moon) is spelled out in separate letters in the right margin of the manuscript.

[251] In Syriac letters in the manuscript. The word قمر ("Moon") is spelled out in separate letters in the right margin of the manuscript. It appears to be an incorrect reading of the Syriac letters in the text.

fol. 25v

Take a part of it and melt it in an iron spoon. Throw ground salt on it and stir it with a piece of iron until it becomes powder. Then take it out, grind it finely, and then put it in a terracotta vessel. Close its head and work on it reinforced clay.[252] Let it spend the night in a bread oven and take it out. Boil it with enough sweet water to cover it and leave it until it settles. Then pour out the salty water and repeat on it the operation with sweet water. Give it a good stir, leave it again until it settles and pour out the water. Do like this with it repeatedly until you do not find in it [any] taste of salt and you do not see any blackness in it. Then dry it up until it dries and becomes calx. Then take as much Yemeni alum, grind it, and dip it in the same amount of sharp vinegar and aged urine. Throw on it the same weight of Egyptian eagle and leave it in this until it dissolves. Cerate the calx with this water little by little, by grinding it, watering it, and drying it in the sun. Your watering should happen drop by drop and not all at once, until it becomes cerated and moist. The tenth or the eleventh time, you will see that it has already reached softness and stickiness to the point that it will stick to the pestle in the mortar [because] of its softness and its stickiness. Then take one ounce of Venus, melt it in a crucible, and throw on it one *dirham* [of the preparation]. Blow on it until it melts, and the medicine penetrates into it. Then it will become Moon, with the help of God.

[§ 35. Making silver from tin (ii)]
Another method. Ten *dirham* of *qalaʿī* should be taken and melted in an iron spoon and a *dirham* of the Moon should be thrown on it with a *dirham* of stone.

[252] The text prescribes shaping a layer of clay around the head of the vessel. The adjective describing the kind of clay to be employed (*muḥkam*, "solid," "firm," "strengthened") shares the same root with the noun *ḥikma* ("wisdom"). A reference echoing the famous clay of the sages, known in the Latin world as *lutum sapientiae* or *argilla philosophorum* may underlie this passage.

fol. 26r

فإذا التحم رميت عليه درهم زرنيخ مبيض ويرمى الثلاثة دراهم على القلعي المذكور فهذا يكون إكسير ألق منه درهما [a] على 8 [253] من القلعي يبيضه ولم يفارقه أبدا ولا يمتزج القلعين بالجسد المذكور حتى يمتزج بالزهرة كذلك يرمي على الجسد المذكور فاعلمه [§ 36] **وجه آخر** خذ خلا حاذقا تحل فيه ملحا وعقابا فإذا انحل فخذ من القلعي ما شئت واسبكه في مغرفة حديد فإذا ذاب تصب عليه القطران وتركته حتى يحترق ثم صبه في الماء المذكور افعل به مرارا حتى يرضيك لونه وإن شئت أذبته وصببت عليه من هذا الماء حتى يتفتت فاسحقه به حتى يشربه فافعل به بأيهما شئت تصل بعون الله إلى الصنعة التي كتمها الفلاسفة إن فهمت وسهل الله عليك وقيل الكيمياء هو الأسرب وقيل هو الشمس [§ 37] **القول في الأسرب** [254] وهو بارد يابس هو من قسمة [b] كيوان وفيه القمر والشمس بالإمكان لا بالعيان وهو ثقيل الجرم بطيء [c] الحركة قابل الصبغ فإذا قبله لم يفارقه إذا أقمته يخرج قمرا وهو يبتها ويصنع منه مرتكا وإسفيداجا وأسرنجا ومنه يكون الإكسير الأعظم الحمرة والبياض ويصنع منه المياه فيحبس بها العبد وماؤه ملائم [d] المريخ وإذا مازجه لم يفارقه ويمتزج بالقلعي ولم يفارقه ويصلح مع الزهرة مدبرا ويلتحم مع الفضة إلا أنه ينفصل عنها

---

[a] درهم [b] قسمت [c] بطي [d] ملايم

---

[253] The figure in the manuscript can be read as the Eastern form of the Arabic numeral 5. See n. 209, above.

[254] A note on the left margin of the manuscript repeats the title of this section, probably with the aim of clarifying the text that is here crammed and unclear.

fol. 26r

When they unite, throw on it a *dirham* of whitened arsenic. Three *dirham* are projected on the mentioned *qalaʿī* and this will be an elixir. Throw one *dirham* of it on five [*dirham*] of *qalaʿī*: it will whiten it, it will never separate from it and the two *qalaʿī* will not mix with the mentioned body until it is mixed with Venus. It is thrown in the same way on the mentioned body. Know it!

[§ 36. Making silver from tin (iii)]
Another method. Take sharp vinegar and dissolve in it salt and eagle. When it dissolves, then take what you wish of *qalaʿī* and melt it in an iron spoon. When it melts, pour the tar[255] over it and leave it until it burns. Then pour it in the aforementioned water. Do [this] with it repeatedly until its colour satisfies you. If you wish, melt it and pour over it [some] of this water until it breaks into pieces. Then pulverise it with it until it has absorbed it. Act on it with any of the two [methods] you wish and, with the help of God, you will reach the art that the philosophers concealed, if you understood and God made it easy for you. It is said that alchemy[256] is lead. And it is [also] said that it is the Sun.

[§ 37. On lead]
Discourse on Lead. It is cold and dry. It is from the part of Saturn, and the Sun and the Moon are in it potentially and not apparently.[257] It has a heavy mass and slowness of movement. It receives the tincture which does not separate from it. If you rectify it, the Moon will come out, which lead stabilises. Litharge, white lead, and red lead are produced from it. The great red and white elixirs are produced from it, and also the waters with which the servant is restrained. Its water is suitable for Mars. When it is mixed with it, it does not separate from it. It mixes with the *qalaʿī* and does not separate [from it]. It is beneficial with Venus and it unites with silver, but it separates from it

[255] For the meaning of "tar" in this context, see n. 217, above.

[256] The word الكيمياء does not always refer to the discipline of alchemy, but has the more general meaning of "a mean to reach an end" (in this case, the elixir). See, for instance the work of the philosopher al-Ġazālī (d. 1111) called *Kīmiyā' al-Saʿāda* ("The alchemy of happiness") where *kīmiyā'* describes a methodology for happiness. In the alchemical context, *al-kīmiyā'* can refer to the philosophers' stone itself, or to its prime matter. Therefore, this passage appears to identify lead as the prime matter required in order to transmute metals into gold. In the Latin tradition, the identification of the prime matter with lead is commonly attributed to al-Rāzī; see, for instance, Rampling, *Experimental Fire*, 116 n.43. This very text could be responsible for such an attribution.

[257] This passage makes reference to the alchemical theory of the hidden qualities of metals. In this case, gold and silver are present in lead *in potentia*, but they are not apparent to the senses. The alchemist's job is to draw out these inner qualities and elicit the transmutation.

fol. 26v

بالخلاص ولا يصلح غبيطا مع الشمس ورائحته [a] تفتت الشمس وتعقد العبد وزعمت فلاسفة الهند أنه أقرب الأجساد إلى الشمس لأن باطن الشمس بارد يابس وباطن الأسرب حار وراطب فثبت أن البارد اليابس من هذه الأجساد هو الأسرب فباطنه بلا شك شمس بخلاف القلعي الذي باطنه قمر وظاهره شمس لأن وجدنا الشمس والقمر في الأسرب أقرب علينا من القلعي وأمكن وقال اعلموا أن الأسرب هو المغنيسيا وأن السر كله فيها وهي المرأة الرخصة وقال أن هذه المرأة الرخصة فيها ٣ أشياء السواد والبياض و الرخص وفيها أيضا ٤ أشياء الرطوبة وسرعة الذوب واللين واليبس وأنها عقرب تحرق وتحترق وفيها البرودة لأنها تطفي حرارة الذكر فلذلك سميت ماء العقرب ولا نعلم شيئا[b] أقوى منها ولا أمكن ولا أقرب منها وقال لا تحسبوا الأسرب مع القلعي ولا مع شيء من الأجساد وما هو إلا شمس دخلت عليه أفة [c] معه في معدنه مغيرة كما تدخل على الجنيب في بطن أمه وهو أخو الشمس في جميع أحواله لأن ✎ [258]خالد وهو خالد والشمس رزين وهو رزين والشمس أبكم وهو أبكم والشمس يصبر تحت الأرض وهو يصبر كذلك فإن أردت [259] محنته وإختباره فأحرقه في ملة بنار معلومة

---

[a] رايحته [b] شياء [c] افت

---

[258] The oblong symbol in the text is interpreted by a superscript note that clarifies it: it should be understood as the Sun and, therefore, as gold.

[259] This verb written below the line and its positioning in the text is indicated by a pointer.

fol. 26v

by means of purification. It is not content with the Sun.[260] Its exhalation weakens the Sun and congeals the servant. The Indian philosophers maintained that it is the closest body to the Sun, since the interior of the Sun is cold and dry and the interior of lead is hot and humid.[261] It is proven that the cold and dry among these bodies is lead and there is no doubt that its interior is the Sun, as opposed to the *qalaʿī*, whose interior is Moon and exterior is Sun, because we found that the Sun and the Moon for us are closer and firmer in lead than [in] the *qalaʿī*. He said:[262] Know that lead is the magnesia and that the whole secret lies in it. It is the tender woman. He said that this tender woman has within herself three things: blackness, whiteness, and tenderness; and it has also four things: humidity, quickness in melting, softness, and dryness. [He said] that it is a scorpion that burns and is burnt and that there is coldness in it because it brings to the surface the hotness of the male. For this reason, it was called the water of the scorpion and we do not know anything stronger nor firmer nor closer than lead. He said:[263] do not count lead with the *qalaʿī* nor with any other body: it is nothing but Sun in which a blemish entered that was with it in its mineral, transforming [it] like [a blemish] enters the foetus in its mother's womb. It is the brother of the Sun in every aspect, since the Sun is immortal and [lead] is immortal, the Sun is grave and it is grave, the Sun is mute and it is mute, the Sun endures under the earth and so does it.[264] If you desire to assay it and test it, then burn it in embers with a certain fire

[260] According to their qualitative nature, lead and gold are the mirror images of one another, and, therefore, they would not naturally mix together. It is also true that lead and gold do not form alloys together.

[261] This passage provides another clear example of the theory of the hidden qualities, establishing a very close relationship between two metals in which hidden and apparent qualities are the exact opposite.

[262] While the Arabic text does not provide an explicit attribution for the following passage, both the Latin editions (Ruska ed. § 58; Steele ed. § 68) and the Hebrew translation (§ 58) attribute this saying to Pythagoras. On Pythagoras as alchemist in the Arabic tradition, see Ullmann, *Die Natur- und Geheimwissenschaften*, 152–3.

[263] The Arabic text does not name the authority of this saying, while in the Latin edition by Ruska (§ 58a) the quotation is attributed to a certain Anfidrius, which is very likely to be the same authority that in the Hebrew translation of the *OAS* is rendered with a transliteration of the Latin (or rather Italian) name Isidoro.

[264] The passage is clearly referring to the properties of gold, i.e. its incorruptibility, its stability and the fact that it does not produce sound when struck.

fol. 27r

كما يصنع بها أهل أهل [265] الختم [a] فإنه يبيض ويصير إسفيداج ثم إن زادت[266] عليه النار صار [267] أسرنجا [§ 38] **صفة تدبيره** خذ منه جزأ [b] فاسبكه في مغرفة حديد وألقه على وزنه عبد واتركه في صلاية وخذ وزن أحدهما ملحا مشويا وحله في مثله خل حاذق [c] وصفه [d] واسحق به الأسرب شيئا [e] بعد شيء بالسحق والتجفيف حتى يسود فتحرقه بالنار حتى يبيض واغسله عن الملح واستعمله إن شاء الله وإن جعلت مع الخل الذي تسقي به مع الملح عقابا جاء غاية والسلام [§ 39] **وجه آخر** خذ من الأسرب والآنك أيهما شئت وأذبه في مغرفة حديد وطاعم كل رطل منه وهو دائرا [f] بأوقيتين من عقرب للأسرب [g] وأوقيتين من علم للآنك [h] وحركه بحديدة تكون معك حتى يصير ترابا أسودا[i] من ساعته ثم اسحقه بوزنه ملحا محلولا في الخل الحاذق [j] فإذا استوى لينه اجعله في قديرة مزججة صابقة على النار وأوقد عليه عشر ساعات حتى يخرج كلس أبيض إن شاء الله [§ 40] **القول في المريخ** وطبعه حار يابس وقيل بارد يابس وهو ذكر وأنثى حامض في الطعم شديد القوة صابر على النار مقاتل لها ويذاب بأربع أشياء بالعلم والأسرب والقلعي والمغنيسيا والمرقشيثا ومتى علق الشمس بالحديد لم يفارقه أبدا ولا يخلص ولا يغيره وإذا خالط الشمس مع القمر صبغ القمر حسنا وأقامها للخلاص وهومن أسرار

---

[a] الحتم [b] جزوا [c] حادق [d] صفيه [e] شيء [f] داير [g] الاسرب [h] انك [i] اسود [j] الحادق

---

265 The word اهل is duplicated in the manuscript.
266 The final ت of this verb is superscripted to the line of text.
267 The verb is superscripted to the line of the text.

fol. 27r

as the people of the seal do with it.[268] It whitens and becomes white lead. Then if you increase the fire on it, it becomes red lead.

[§ 38. Preparation of lead (i)]
Description of its operation. Take a part of it and melt it in an iron spoon. Throw it on the same amount of servant and leave it in a mortar. Take the weight of one of them of roasted salt, dissolve it in the same amount of sharp vinegar, and filter it. Grind the lead with it little by little by grinding and drying until it blackens. Then burn it with fire until it whitens and wash it from the salt. Use it, if God wills. And if you put [some] eagle with the vinegar with which you water the salt, a higher goal and peace derives from it.

[§ 39. Preparation of lead (ii)]
Another method. Take either lead or tin, as you wish, melt it in an iron spoon, and, while it rotates, feed every *raṭl* of it with two ounces of scorpion for lead or two ounces of the sign for tin. Stir it with a piece of iron that you have, until it suddenly becomes a black soil. Then grind it with the same weight of dissolved salt in sharp vinegar. When its softness becomes uniform, put it in a fire-resistant glazed pot and light the fire on it for ten hours until a white calx will comes out, if God wills.

[§ 40. On iron]
Discourse on Mars. Its nature is hot and dry. Some say cold and dry. It is male and female, sour to the taste, very strong and resistant to the fire which it fights. It is melted with four[269] things: with the sign, lead, *qalaʿī*, magnesia, and marcasite. When the Sun is attached to iron it never separates nor detaches from it and does not change it. If the Sun is united with the Moon, it dyes the Moon well and through purification rectifies it. It is among the secrets

[268] The phrase "people of the seal" may point to artisanal practices of melting lead to produce seals for scribes and official correspondence.
[269] Although the text specifies four things, five are actually listed.

fol. 27v

الهند مرادهم المريخ وحده من سائر [a] الأجساد وكذلك اختارته الحكماء لأنه هين النقض سهل التركيب وهو ثابت الصبغ في العمل في البياض والحمرة وتدبيره على ضربين أحدهما إذابته بالشب وإذابته بالعقاب وسوف نذكر ذلك جميعا [§ 41] صفة **تدبيره** قد ذكرنا أن المريخ أشد صلابة من جميع الأجساد وهو لا يذوب في النار اللينة إلا بعلاج شديد وذلك أن تأخذ من صفائحه[b] الرقاق فتحميها وتطفيها في زيت قد أذيب فيه أسرب مرارا حتى يتفتت ويلين جدا ثم قطعها بمقراض أرق ما تقدر عليه ثم خذ من الأسرب جزأ[c] أذبه بمغرفة حديد وألق عليه وزنه العبد ثم اسحقه حتى يصير غبارا فاجعل منه طاقا ومن الصفائح [b] طاقا حتى تملاء البوط ثم ركبه في الفرن وانفخ عليه بنار الكير حتى يذوب ثم أفرغه وعاود عليه مرارا حتى يرضيك لونه ويسرع سبكه ثم استعمله فيما تريد [§ 42] **وجه آخر** خذ من برادة المريخ ما شئت وألق[d] عليها من العلم الطيب المسحوق مثل ربعه واجعل ذلك في صرة دون دهن وطين عليها بطين الحكمة فإذا جفت اجعلها في فرن السبك وانفخ عليها بالكير ٣ ساعات ثم تردها وتخرجها واسحقها مع سدسها ملح قلي ملتوت واجعلها في بوط بر بوط *حتى ينزل أسفل البوط *[270] وانفخ عليها بالكير

[a] ساير [b] صفايح [c] جزوا [d] القى

[270] The phrase between brackets is added to the lower margin of the manuscript probably with the function of catch-word, and is also repeated at the beginning of the next folio, where is it seems to find its correct positioning.

fol. 27v

of the Indians [and] one of their goals. Mars is unique among the rest of the bodies. Similarly, the sages chose it because it is simple to destroy and easy to prepare; its tincture is stable in the work of the whiteness and of the redness. Its preparation is of two varieties: melting it with alum and melting it with the eagle. We are going to mention all of this.

[§ 41. Preparation of iron (i)]
Description of its operation. We have already mentioned that Mars is the hardest of all the bodies and that it does not melt in a gentle fire without a strong treatment. This is, that you take thin sheets of it, you warm them and temper them repeatedly in oil, in which lead has been melted several times, until they are broken up and well softened. Then cut them in pieces as fine as possible with scissors. Then take a part of lead and melt it in an iron spoon. Throw on it the same amount of the servant and then grind it until it becomes powder. Place a layer of it and a layer of the sheets until the crucible is full. Then insert it into the oven and blow on the fire with the bellows until it melts. Then empty it and repeat the operation several times until its colour satisfies you and its melting is fast. Then use it for anything you want.

[§ 42. Preparation of iron (ii)]
Another method. Take as much as you want of filings of Mars and throw onto them a quarter of its amount of good ground sign. Put them in a bag without fat and cover it with the clay of wisdom. When it dries, put it in an oven for melting and blow on it with the bellows for three hours. Then bring it back, take it out [of the bag], and grind it with a sixth [of that amount] of ground alkali salt. Put them in a descensory[271] *until it descends to the bottom of the crucible* and blow on it with the bellows

[271] The descensory is a particular kind of composite crucible which was known in the Latin world as the *botus barbatus* or *descensorium*. On distillation *per descensum* and its apparatus, see Thomas and Claude, "Vases à fond percé."

fol. 28r

حتى ينزل أسفل البوط ويبقى الخبث في البوط الأعلى ثم أخرج ما نزل فرده إلى الذوب فإن شئت اخذت زجاجا وتنكرا واسحقهما بالزيت واجعلهما بنادق وطاعمته منها وأنت تذيبه أبدا مرارا فإنه يزداد سرعة ذوبه وبياضه وإن اخترت من ذلك لان حتى يذوب في مدة يسيرة مثل ذوب الأسرب بالشب وإذابته بالعقاب وسوف نذكر ذلك جميعا إ شاء الله [§ 43] وجه **آخر** يؤخذ ܡܢ ܡܐ[272] القلي ما شئت فاعجن به العلم الأصفر والطخ به صفائح [a] المريخ وأحمها [b] وأطفها [c] في ذلك مرارا حتى تبيض وتصير بلون القمر سواء [d] فاعزله فإنه صلب لا فرق بينه وبين القمر إلا صلابة فيه فقط فاسحقه وألقه [e] في بوط وألق عليه قطعة من المرتك المبيض أو من الإسفيداج أو قطعة من بورق الحكماء أو ما يجري هذا المجري فإنه ينحل سريعا مثل الماء فإنه انحل فأنت فيه بالخيار إن شئت تفرغه الآن وإن شئت أعد عليه السبك مرارا حتى يرضيك لونه فإنه يصير في قوام الأسرب في الذوب ثم امزجه بالقلعي وامزجهما أيضا بالقمر واعمل منهما ما شئت فإنه سريع العمل قريب المدة إلا أن فيه هذا الزمن في وقت تليين المريخ فقط [§ 44] **غيره آخر** ملح القلي المبيض المسحوق المستخرج وبول

[a] صفايح [b] احميها [c] اطفيها [d] شوا [e] القيه

[272] The previous two words are written in Syriac letters in the manuscript. A note on the left margin of the manuscript provides a transcription in Arabic letters: من ماء .

fol. 28r

until it descends to the bottom of the crucible, leaving the dross in the upper crucible. Then take out what descended and bring it again to melt. If you want, you [can] take glass and borax, grind them with the oil, shape them in [the shape of] hazelnuts and feed it with them, while you continuously melt it several times. This will increase the quickness of its melting and its whiteness. If you prefer, instead of this [method], soften it until it melts in a short span of time,[273] similar to how lead melts with the alum and how it is melted with the eagle. We will mention all of this, if God wills.

[§ 43. Preparation of iron (iii)]
Another method. As much alkali water as you want should be taken. Knead with it the yellow sign and smear sheets of Mars with it. Warm them and temper them in this several times until it whitens and assumes the colour of the Moon evenly. Then separate it: it will be hard and there will be no difference between it and the Moon apart from its hardness alone. Then grind it, throw it in a crucible, and throw on it a piece of whitened litharge or white lead, or a piece of borax of the sages or of what goes in this direction. It will dissolve quickly like water and you will have a choice: if you want, finish it straight away or, if you want, repeat the melting on it many times until its colour satisfies you. As for melting, it will reach the same condition as lead. Then mix it with the *qalaʿī* and mix the two also with the Moon. Do what you want with them. It will be quick to work upon in a short time, not considering the time required for softening Mars and nothing else.

[§ 44. Preparation of iron (iv)]
A different [preparation]. Whitened ground extracted alkali salt and urine

[273] The reading of the manuscript is problematic in this passage; my proposed translation is based on the other versions of the *OAS* and the context, and should be considered provisional.

fol. 28v

الغزلان خذ ما شئت منه وخذ صفائح[a] المريخ ورقها وأطفها [b] في الماء المذكور عشر مرات [c] حتى تبيض الصفائح [a] وتترطب فاقطعها بمقراض وطبقها في البوط مع إسفيداج الأسرب الملتوت بالزيت واسبكه يخرج ☾ [274] أغبر اجعل [275] عليها العظم والأسرب يخرج لك نصفها ☾ طيبة فزاوج كيف شئت فإنه صحيح مجرب [§ 45] **آخر** خذ ٤٠٠ ܡܢ [276] العقرب الأصفر جزأين [277] [d] فاسحقه ثم اجعله في شقف وأغله [e] في مثله زيت حتى يذوب العقرب فيه ثم خذ صفائح [a] المريخ أربع أجزأ فأحمها [f] وأطفها [g] في ذلك الزيت مرة بعد مرة حتى تحترق الصفائح [h] وتسود وتسحق [§ 46] **غيره** خذ من برادة المريخ جزأ [i] ورشها بخل حاذق [j] وضعها في الشمس ثم عاودها مرارا حتى تراها مثل الزنجفر والزعفران فاسحقها بسدسها عقربا وأحرقها في فرن الخبز ثم اسحقها بماء الزاج والعقاب كما تقدم وألق [k] منها درهم على ٤ من القلعي أو القمر تخرج ٣٠٠ ܡܣ [278] إن شاء الله تعالى وإن شئت أخذت من الصفائح [h] وغطتها[l] بالعقرب المذاب بالزيت وركبتها في قديرة مزججة وغطتها[l] مشدود وصلها بطين الحكمة وبيتها في فرن خبز أو في قصني وهو زبل الخيل ثم أخرجها واسحقها

---

[a] صفايح [b] اطفيها [c] مراة [d] جزوا [e] اغليه [f] فاحميها [g] اطفيها [h] الصفايح [i] جزوا [j] حادق [k] القى [l] غطيتها

---

[274] The symbol of a crescent here stands for the product of this operation, which is silver.

[275] This verb is superscripted to the line of text.

[276] An Arabic numeral (400) is here followed by the preposition ܡܢ in Syriac letters. A marginal note repeats the same preposition in Arabic script.

[277] The Arabic letter *nūn* is superscripted to the line of text, possibly indicating the dualization of this noun.

[278] An Arabic numeral (400) and two Syriac letters are combined to indicate the product of this operation. 400 corresponds to the Arabic letter ش and the two Syriac letters correspond to م and س, thus the encoded word can be decoded as "The Sun" (شمس) and hence as gold.

fol. 28v

of the two gazelles; take what you want of it.[279] Take the sheets of Mars, make them thin, and temper them ten times in the abovementioned water until the sheets whiten and become moist. Then cut them with scissors, stratify them in a crucible with lead white [made] from lead, ground with oil,[280] and melt it. Ash-grey ☾[281] will come out. Put in it bone and lead. Half of its amount of fine ☾ will come out for you. Combine it then as you want, since it is good and tested.

[§ 45. Preparation of yellow sulphur]
Another [operation]. Take [400][282] two parts of yellow scorpion, grind it and then put it in a pot sherd. Boil it in the same amount of oil until the scorpion dissolves in it. Then take four parts of the sheets of Mars, heat them and temper them in that oil time after time until the sheets burn, blacken, and are ground.

[§ 46. Preparation of iron (v)]
Another [operation]. Take one part of filings of Mars, sprinkle them with sharp vinegar, and place them in the sun. Then repeat [the operation] on them several times until you see them like cinnabar and saffron. Grind them with a sixth [of their amount] of scorpion and burn them in a bread oven. Then grind them with water of vitriol and eagle as before. Throw one *dirham* of it on four [*dirham*] of *qalaʿī* or Moon. Sun will come out, if God, may He be exalted, wills. If you want, take some of the sheets, cover them with sulphur melted with oil, and place them in a glazed pot. Cover it tightly with a joint of clay of wisdom, and leave it overnight in a bread oven or in *qiṣnī,* which is horse manure. Then take them out and grind them

279 Urine is a very common organic ingredient in alchemical recipes that often specify which kind of urine should be employed; recipes prescribe the use of boys' urine, young girls' urine, and aged or matured urine. This recipe possibly requires the use of a mixture of female and male gazelles' urine. The name "gazelle" could also be a rare *Deckname* standing for another substance.

280 This passage suggests that the author believes lead white could be produced starting from substances other than lead.

281 The symbol of a crescent here stands for the product of this operation, namely silver.

282 The Arabic numeral 400 is found in this position in the manuscript, followed by the Arabic preposition من in the Syriac alphabet. The role of these figures in this sentence remains obscure, since the quantity of yellow sulphur required in the recipes is expressed as "two parts" in what follows.

fol. 29r

سحقا بالغا ثم خذ من الزاج وزن المريخ فأشويه في فرن الخبز ساعة وحله في مثله أبوال المهى المعتقة في زجاجة ثم ألق فيه وزنه من العقاب الأبيض المصري الطيب واتركه حتى ينحل وشمع به الجسد شيئا [a] بعد شيء حتى يستوعبه ثم ألق منه درهما على أوقية من الأسرب يخرج الشمس [283] [§ 47] **القول في الزهرة** وطبعه الحرارة واليبس دون يبس المريخ وقوته في الذوب والصلابة مثل قوة القمر غير أنه أحمر من قلع حمرته عاد قمرا وهو في الظاهر زهرة والباطن قمر نفيس وهي أخت القمر وتمتزج مع الشمس وتقبل الصبغ ويكون منها إكسير البياض والحمرة ولا يثبت صبغه إلا بماء الزاج وبماء الشب وماء الزهرة يذيب الأجساد وكل شيء من الحجارة وتدبيره كتدبير المريخ سواء[b] [§ 48] **صفة تدبيره** خذ من صفائح [c] ما شئت فرققها واغمسها في زيت قد أذبت فيه شيء من العقرب أو من علم ويكون العقرب سدسها فأطفها [e] به نعما وركبها في كوز وافرش عليها ملحا مسحوقا واعمل منه طاقا وأعد عليها الملح حتى تكملها ثم أطبق عليها وشد الوصل وبيتها في فرن الخبز فإذا أخرجتها من الغد سحقت الصفائح [c] في هاون بنحاس

---

[a] شي [b] سوى [c] صفايح [d] فاطفيها

---

[283] The word شمس (Sun) is spelled in the Syriac alphabet in the manuscript.

fol. 29r

finely. Then take of vitriol the same weight as Mars, roast it in a bread oven for one hour, and dissolve it in the same amount of antelope urine aged in a glass bottle. Then throw into it the same weight of good white Egyptian eagle and leave it until [it] dissolves. Cerate the body with it little by little until it holds the mixture. Then throw a *dirham* of it on an ounce of lead. The Sun will come out.

[§ 47. On copper]
Discourse on Venus. Its nature is hotness and dryness, without the dryness of Mars. Its faculty to melt and [be crushed in] the mortar is like the faculty of the Moon, but it is red. He who extirpates its redness, turns it into Moon. On the exterior it is Venus and on the interior it is the precious Moon. It is the sister of the Moon, it mixes with the Sun and receives the tincture. The elixir of the whiteness and of the redness derives from it. Its tincture is stabilised only with the water of vitriol and the water of alum. The water of Venus melts the bodies and every stone. Its operation is the same as the operation of Mars.

[§ 48. Preparation of copper (i)]
Description of its operation Take as many sheets as you want, then make them thin and dip them in oil in which you melted some scorpion or some sign. Let the scorpion be a sixth of it. Temper them well with it and put them in a clay jug. Spread ground salt over them and make with it a layer. Renew the salt over them until they are complete. Then close the jug, make a good seal, and leave it overnight in a bread oven. Then, when you extract them the day after, grind the sheets with copper in a mortar

fol. 29v

سحقا ناعما ثم اغسلها عن الملح بالماء العذب حتى لا يبقى للملح فيه طعم فجففه وارفعه فهو الزهرة الذي لا ظل لها فإن أردت الحمرة فخذ من الزاج المشوي في الفرن مثل وزن الزهرة فاغمره بمثله من الماء العتيق واتركه فيه حتى ينحل ثم صفه تصفية حسنة وألق [a] عليه مثله عقاب فإذا انحل فشمع به الزهرة المحرقة شيئا [b] بعد شيء حتى يستوعبه ويصير في لون الزنجفر فخذ منه جزأ [c] ومن الأسرب جزأ [c] فالحفه به في بوط و واسبكه حتى يغوص فيه يخرج ⬭ [284] وإن شئت أخذت منه جزأ [c] ومن القمر جزئين [d] وصنعت منها صفائحا [e] رقاقا ثم افرش تحت الدواء وفوقه في ‡ [285] ثم اسبكها بعد ذلك جيدا وزنها فإن لم تزد في وزنها شيئا [f] وخرج لونها أصفر فاسبكها بعد ذلك ٣ مرات وصبها أبدا في بول غزال معتق حتى تصفو ويخرج وجهها ومحكها أصفر فاجعل منه جزأ [g] ومن الشمس الطيب جزأ [g] و نصف جزء يخرج غاية [§ 49] ܐܚܪܢ [286] خذ من الراسخت ? [287] أجزأ فاسحقها بليغا في مهراس نحاس مع جزء من العقرب واجعل الجميع في قدر مزجج وطينه وألق [h] عليه غطاء محكما وألقه [i] في فرن الخبز ٣ ساعات وإن شئت جعلت موضع الراسخت [j] زنجارا طيبا واغرز فيه العقرب كما ذكرت لك واسحقه بماء الزاج والعقاب

[a] القى [b] شيء [c] جزو [d] جزوين [e] صفايح [f] سيء [g] جزو [h] القى [i] القيه [j] الرسخت

[284] The text introduces here an oblong symbol representing the product of the previous operation.

[285] A word in Syriac letters, probably indicating a kind of apparatus, is found here in the manuscript. I have not been able to decipher it because of the unclear script.

[286] The word spelled here in Syriac letters can be read as the Arabic اخر ("another").

[287] The figure in the manuscript can be read as the Eastern form of the Arabic numeral 5. On the different forms of Arabic numerals, see above, note 209.

fol. 29v

very finely. Then wash them from the salt with sweet water until no taste of salt remains in it. Then dry it and take it up. It is Venus that has no shadow. If you want the redness, then take the same weight of vitriol roasted in the oven as of Venus, submerge it with the same amount of aged water, and leave it there until it dissolves. Then drain it well and throw on it the same amount of eagle. When it dissolves, then cerate with it the burnt Venus little by little until it holds it and becomes the colour of cinnabar. Take a part of it and a part of lead, cover it with it in a crucible and melt it until is absorbed. The Sun will come out. If you want, take a part of it and two parts of the Moon, make thin sheets of them, and spread the medicine under them and over them in the ‡.[288] Then, melt them well; after that weigh them well. If there is no increase in their weight and if their colour comes out yellow, then melt them after this three times and pour them every time into aged gazelle urine until they are purified and their appearance and trial come out yellow. Put a part of it and a part and a half of the good Sun. The highest goal will come out.

[§ 49. Preparation of copper (ii)]
Another. Take five parts of burnt copper, grind them finely in a copper mortar with a part of the scorpion. Put everything in a glazed pot and cover it with clay. Place on it a reinforced lid and put the pot in a bread oven for three hours. If you want, put good verdigris instead of the burnt copper and stick the scorpion in it as I have already mentioned to you. Grind it with water of vitriol and the eagle,

[288] A word in Syriac letter, probably indicating a kind of apparatus, is found here in the manuscript. I have not been able to decipher it.

fol. 30r

كما تعلم وألق منه درهما على ٢ ☾ [289] يخرج ٨ [290] *شمس* فإن أرضاك لونها وإلا اذبها وألق عليها درهم آخر من الدواء وأفرغها واحمد الله تعالى فإن ذلك كله قريب من يوم واحد وأقل وإن أردت [a] مزجه فاجعل منه جزأ [b] ومن الشمس الخالص جزأ [b] فإنه غاية [§ 50] **آخر** خذ من برادة الزهرة جزأ [b] ومن العبد جزأ [b] فاسحقهما حتى يتمزجا و اغسلها بالماء والملح حتى تنقا ثم اغسلهما بالماء العذب وجففها وخذ من الزاج مثل الزهرة فحله في مثله من بول الغزلان المعتق [291] وصفه [c] وألق فيه وزنه عقابا وشمع بهذا الماء شيئا [d] بعد شيء تشربه ويصير فرفيريا قد زاوج بعضها بعض فاستعمله تسر به إن شاء الله وإن أردت للبياض جعلت مكان الزاج شب يماني وصنعته كما تقدم وطرحت منه على الزهرة يخرج قمرا إن شاء الله تعالى [§ 51] القول في الـ*ز ج ا ج* [292] وهو من قسمة [e] عطارد الغالب على طبعه البرد واليبس إلا أنهم يعملوه من قسمة [e] عطارد ولا اعرف من أين وجه ذلك لأنه يتلون بأنواع الألوان ويقبل الأصباغ ويذيب المريخ وجميع الأجساد ويجف بها في السبك ولم اجرب منه أكثر ما أنا اذكر ما جربت على بركة [f] الله وعونه [§ 52] **صفة** ‡ ا ܣܡܩܐ[293] خذ من الزجاج العراقي واسبكه في بودقة فإذا ذاب طرحت عليه حبة من شاذنة

---

[a] ردت [b] جزو [c] صفيه [d] شيء [e] قسمت [f] بركت

---

[289] Over the symbol of the crescent a superscript note clarifies that it should be read as the Moon/silver.

[290] A upside down heart shaped symbol represents here the product of the previous operation. A superscript note clarifies that it should be read as the Sun/gold.

[291] A subscript ع is added to clarify the reading of this word, which is written in a rather crammed fashion in the manuscript.

[292] The title of this section of the manuscript is spelled in Syriac letters in very unclear handwriting, and its reading would have been very difficult, were it not for a marginal note that clarifies, in detached Arabic letters, that the Syriac letters correspond to the word زجاج ("glass").

[293] The title of this section of the manuscript is spelled in Syriac letters and its reading is very uncertain. The second element can be read as "red" and the context and comparison with the Hebrew manuscript and the Latin editions suggest that the section deals with a red gem or, more generally, with a red precious stone.

fol. 30r

as you know, and throw one *dirham* of it on two [*dirham*] of the Moon. Sun will come out. If its colour satisfies you [this is good], otherwise melt it and throw on it another *dirham* of the medicine. Pour it out and praise God, may He be exalted. All of this [takes] close to only one day or less. If you wish to mix it, put one part of it with one part of pure Sun: it will be a [highest] goal.

[§ 50. Making silver with copper and mercury]
Another. Take a part of filings of Venus and a part of the servant, grind them until they are mixed, and wash them with water and salt until they are clean. Then wash it with sweet water, dry it, and take as much vitriol as Venus. Then dissolve it with the same amount of aged urine of the two gazelles and purify it. Throw in it the same weight of the eagle and cerate [it] with this water little by little. It will absorb it and become purple. They will already be united with each other. Then use it. Rejoice with it, if God wills. If you wish [to obtain] the whiteness, put Yemeni alum instead of vitriol, and work it as already mentioned. Throw some of it onto Venus: it will come out as the Moon, if God, may He be exalted, wills.

[§ 51. On glass]
Discourse on glass.[294] It is from the part of Mercury. The predominant [feature] of its nature is cold and dry, except that they worked out that it is from the part of Mercury (but I do not know where this opinion comes from) since it is coloured in a variety of colours, receives the dyes, melts Mars and all the bodies, and dries with them in the melting.[295] I did not try with it anything more than what I mentioned I have tried, with the blessing and the help of God.

[§ 52. Preparation of artificial ruby]
Description the red gemstone.[296] Take some Iraqi glass and melt it in a crucible. If it melts, throw on it a grain of prepared

[294] The title of this section of the manuscript is spelled in Syriac letters in very unclear handwriting, and its reading would have been very difficult, were it not for a marginal note that clarifies, in detached Arabic letters, that the Syriac letters correspond to the word زجاج ("glass").

[295] The general reference to "they" as the authorities behind this opinion should be understood as a general mention of the alchemists or, more extensively, the philosophers.

[296] The reading of the Syriac word in the title of this section is reconstructed on the basis of the general structure of the treatise and the content of the following passages. The Syriac text is difficult to decipher.

fol. 30v

مدبرة ثم افرغ منه على اي قالب شئت واتركه يبرد يأتيك من أجل ٣٠٠٦١٠٠١١٠٣٠١ [297] [§ 53] **صفة ٧٠ص اصى؛** [298] خذ من الزجاج العراقي الطيب ما شئت فاسبكه بشيء من ملح القلي فإذا ذاب [a] اقلب فيه وزن حبة من زعفران الحديد المدبر بماء الزاج فإن أرضاك لونه وإلا رد عليه من الإلقاء والسبك حتى يرضيك يخرج أصفر كلون الياقوت [§ 54] **صفة فص فيروزج** أذب من الزجاج العراقي كما تقدم ذكره وألق [b] فيه ربع حبة مغنيسيا مدبرة باللازورد المعدني المحلول بماء الشب و أسقه [c] شيئا [d] بعد شيء فإن أتاك لونه غمائما حسنا وإلا عد عليه حتى يرضيك لونه [§ 55] **صفة ٤٢٠٠٤٠٧** [299] خذ من الزجاج على الصفة وألق عليه حبة من زنجار ودهنج محلولين شيئا [d] بعد شيء حتى يرضيك لونه إن شاء الله تعالى انتهى قولنا على الأجساد والأرواح وبالله المستعان

---

[a] داب [b] القى [c] اسقيه [d] شيء

---

[297] The word *al-yāqūt* ("ruby") is hidden behind a series of Arabic numerals, a cipher that uses the correspondence between numbers and letters of the *abjad system*. A superscript note in separated Arabic letters clarifies the meaning of the phrase.

[298] The title of this section is encoded with a mixture of Arabic numerals, Arabic letters, and Syriac letters. A superscript note provides its interpretation in separate Arabic letters: "description of the yellow gemstone."

[299] The word *zumurrud* ("emerald") is hidden behind a cipher that relies on the correspondence between numbers and letters of the *abjad system*. A superscript note in Arabic clarifies the meaning of the phrase.

fol. 30v

haematite. Then pour some of it in any mould you want and leave it to cool. It will give you the best precious stone.

[§ 53. Preparation of an artificial yellow gemstone]
Description of the [artificial] yellow gemstone. Take as much as you want of good Iraqi glass and melt it with some alkali salt. When it melts, pour over it the weight of one grain of saffron of iron prepared with water of vitriol. If its colour satisfies you, [this is good], otherwise renew the pouring and melting until it satisfies you. It will come out yellow like the colour of a precious stone.

[§ 54. Preparation of an artificial azure gemstone]
Description of the turquoise. Melt some Iraqi glass as we have already mentioned and throw into it a quarter of a grain of magnesia prepared with mineral lapis lazuli dissolved in water of alum. Water it little by little. If it gives you a colour beautiful like the clouds, [this is good], otherwise repeat [the operation] on it, until its colour satisfies you.

[§ 55. Preparation of an artificial green gemstone]
Description of emerald. Take some glass according to the description and throw on it a grain of verdigris and of malachite, both dissolved, little by little until its colour satisfies you, if God, may He be exalted, wills.

Here ends our treatise on the bodies and the spirits and God is the only one whose help can be sought.

MS Orient. Oct. (klein) 514 – Berlin Staatsbibliothek

fol. 19v

22 אתחיל ספר אלומי ומלחים

המאמר באלומי ובמלחים הצריכים למלאכה הזות. [§ 1] וזהו טבע

fol. 20r

האטרמנטו. דע כי האטרמנטו הם מינים רבים ומנירהם[a] כן נמצא. והם מים וצבע. שהקפיאו יובש העפר. וטבעם חם ולח. ואחד ממיניו הוא קלקדור ושורין וקלקדים וקלקנת. והטוב מכלם הוא אצלנו באיי אספנייא שמביאים אותו מסבלא. ואלו האטרמנטא משחירים הגופות נ״ל אם לא יודלפו. ויוסיפו אדמימות לאדום. וישחירו הלבנות והיותר דק ממנו הוא קלקנתי. והיותר עב הוא שורין. ואמר ג׳יבר בן חיין בהפרח המדבר בעניינים המוסטטין אמר דע כי האטרמנטו הוא אוריגי״ ר״ל צמח מאוריגני ר״ל מצמחי המלאכה. וזה הצמח מאד נתעלה אלוסואי קונדימנטו הוא עב אלוקפק. נ״ל עוקב. וזה יעשה אותו פילטריטיבו כי האטרמנטו הוא הדיר אפלטריזיאוני. ואמר כי האטרמנט״ יש גפריתים דקים יעשו דמות ויאירו ובלעז אלו מצנו. ושהוא *נ״ל שמא* יצבעו ותנסה שלא תסתפק ועל כן פתח לבך ונסה לפי שאמר שמא. ואחד מן הסגולות שלו הוא שאותו אטרמנטו הוא שיכתוש כל הבורחים שלא יברחו דע זה. וכלל איכות מלאכתו הוא שתקח ממנו כרצונך ושים ממנו לילה א״ בתנור חם. ואז יתאדם הזאג ההוא אדממות חזק. ואז יעשה אינקורפוררי עמו מים חם ד״ כמוהו. ותנוחנו עד שיותר וימס. ואז תסננהו ושים למשמרת עד שתצטרך.

[§ 2] המאמר רייזוני של אלומי

אלומי הוא אוניינטו שייבשהו יובש הארץ אליו ומערכיו רבים.

ואמר בן יוליאו דקורטובא כי מעדיו *נ״ל מעדנו* אצל חצי היום. ר״ל דרום בארץ קורטובא הנקרא אנאמריים טבעו חם ולח. ואמרו אחרים כי הוא חם ויבש. וזהו הכובש. נ״ל פוחקים. ומנקה הגופות. וינקם נקוי נאה. ויעשה הגופות זכים. ויוסיף בצבעם. אבל הלבן יעשה שחור עד שישקרופה הדמות של טבעו. ואח״כ לא ישתנה. והלבן יעתיך לאדום והוא יכפיל האחר וישלש השנים וירבע השלשה

---

[a] ומנירכם

MS Orient. Oct. (klein) 514 – Berlin Staatsbibliothek

fol. 19v

I begin the book on Alums and Salts. **The discourse** on alums and salts that are needed in this operation.[300]

[§ 1. On atrament]
This is the nature of

fol. 20r

atrament. Know that there are many kinds of *atrament* and that its *mines* are [easily] found. They are [made up of] water and colour that the dryness of the earth has congealed. Their nature is hot and humid. One of their kinds is the *qolqodor*, the *sorin*, the *qalqadim*, and the *qalqant*.[301] The best of all of them is with us in the island of *Spain* that they know as *Sevilla*. These *atraments* blacken the bodies – in my opinion: if they are not distilled – they increase the redness of the colour red and they blacken the whiteness. The finest among them is the *qalqanti* and the thickest is the *sorin*. Ğābir ibn Ḥayyān in the *Flowering of the Desert*[302] about the topic of the mutations said: 'Know that *atrament* is the *origin* – this means: plant – of the *origins* – this means: plants – of the operation. This plant is highly exalted. *Its operation* is *Alwaqfaq* – in my opinion: eagle[303] – and this makes it *permeable* since *atrament* is permeable [to] *infiltrations*. And he said that *atrament* is [made up] of fine sulphurs that constitute its appearance and enlighten it. We found this is a foreign language[304] and that *in my opinion: here* they dye. Try if you are not persuaded: then open your heart and try according to what is said here. Then, there is among its virtues that this *atrament* crushes all the fleeing substances that do not flee. Know this! All the modalities of its operation [imply] that you take as much as you want of it and that you put it for one night in a hot oven. In this way this vitriol[305] will redden with a strong redness. In this way you can *incorporate* it with three times [its weight] of hot water. Let it rest until it is dissolved and liquified. Then filter it and save it until you need it.

[§ 2. On alums]

The discourse on the *method* of alum.[306] **Alum** is an *oily substance* that the dryness of the earth has dried up. Its qualities are many. Ben Yulio from Cordoba[307] said that its mines *in my opinion: its minerals* are in the midday – this means: South – in the land of Cordoba called *Anamariym*. Its nature is hot and humid. Others said that is it hot and dry. It is the conqueror – in my opinion: frightened – [308] and the cleanser of the bodies, since it makes them clean and beautiful. It makes the bodies pure and adds to their tincture. But the white makes [them] black until the character of its nature is taken out. Moreover, it is not subject to change. The white [alum] melts to red and doubles the one, triples the two, and quadruples the three.

[300] See the *Criteria of Edition and Translation*, p. 48, above.
[301] These three terms are transliterated from Arabic in the text. Their identification is complicated by the rendition in the Hebrew alphabet.
[302] This title does not correspond to any known work attributed to Ğābir ibn Ḥayyān.
[303] In transliteration from the Arabic عقاب . Eagle is a common *Deckname* for sal ammoniac.
[304] On the language of the *Vorlage* of the Hebrew *OAS*, see *Introduction* and in particular § 5.6.
[305] In transliteration from the Arabic زاج in the text.
[306] The text uses the Italian word *razione* in transliteration, which is possibly employed for rendering the Latin *ratio*.
[307] To be possibly identified with Ibn Ğulğul of Cordoba (d. 994); see *Introduction*, § 1.2 and § 4.3.
[308] The meaning of this word and of the note itself remains obscure to me.

fol. 20v

ואמרו כי אלומי הוא יותר נובילי מכל האבנים כי הוא קר בעבור חמוצו והוא יותר קונבינינטו מכל הדברים לתקן הזרניך. נ״ל גם הגפרית איסינבליצימינטי יותר קונסליבילי הוא זה. ואיכות מלאכתו שתקח אלומי ימיני אותו שהוא לבן אילמינוסו איפורו. ויכתשוהו היטב ושים בקדרה מזוככת. ושים עלה שתן נערים. ואחר שיתבשל השתן וינוח ויצליל. אחר בשלהו עם אלומי ההוא באש רפה וערב עם תרווד היטב והנח עד שירדו השמרים בתחתית הכלי וסנן היטב ושמר[a] לעת הצורך.

[§ 3] המאמר רייזיני של מלחים

דע כי המלחים הם הרבה מינים. המובחר שבכלם הוא סל יימא אחריו מלח הלחם. ואחריו מלח הנדי והוא אדום ס״א זה המלח החזק. ואחריו מלח הסיד. ואחריו מלח מר. והוא באי[b] אספנייא הנקרא אנדלוס. מהבאציליקון. וטבעו חם ויבש והוא מים שיבשהו חממות האש. ומסגולותיו הוא שהוא לבדו יחזק האש ויוסיף לבנות לכסף. נ״ל לב״ בדילין. ויעתיקנו דלוקורפורלי אנישפוראליטטי לובן יעשה עם הזהב ויוסיף אדמימות וירחץ הגופות מן הטינוף. ועמו קליסן הכל ר״ל אנקלצינטו הגופות לא עם אחר. ועל כן קראוהו הפלוסופים כסף הפילוסופים מפני שהם יצטרכו לו וינהגו עמו בכל דבר והוא הינדימינטו שכל העולם. ועמו שימוליורני כל דבריו הרבה נ״ל בחרו. שנ״א על כל קרבנך תקריב מלח. וכשתדקדק בו אז תמצא גדולתו מלח בכל אפרי בייטיבלי אבן צמחי[c] ובסיד דומם האבנים. ובעצמות האבנים במלחים. ובכל הדברים שימצאו בכל העולם. הוא נסתר פרולסואה סוליויאו

[a] שמור [b] איי [c] צמחיו

fol. 20v

And they said that alum is the *noblest* of all stones, since it is cold on account of its ferment, and that it is the most *convenient* of everything for rectifying *arsenic*[309] – in my opinion: sulphur as well – *and* that this is *simply* the most *advisable*. The modalities of its operation are that you take this Yemenite alum which is white, *shiny and pure* and that you grind it finely; put it in a glazed pot and put on top of it urine of young boys. After the urine has been cooked, it has rested and has settled, then cook it with that alum in a gentle fire and stir it well with a ladle; let it rest until its sediment sinks to the bottom of the vessel. Filter it well and save it until you need it.

[§ 3. On salts]
The discourse on the *method* of salts. **Know** that there are many kinds of salts and that the best among all of them is *rock salt*, then there is bread salt, then Indian salt, which is red – another explanation: this is the strong salt – then the salt of calx and then the bitter salt, and this is found in the island of Spain called Andalus from the *basilicon*. Its nature is hot and dry: it is a water that the hotness of fire has dried. Among its virtues is that it strengthens fire by itself, that it increases the whiteness of silver – in my opinion: of the two [kinds of] tin – and makes it change from *corporality to spirituality*. It whitens with gold, increases its redness and washes the scum from the bodies. With it, and not with anything else, all the bodies are calcined – it means: *calcined*. For this reason, all the philosophers called it silver of the philosophers, since they require it and they use it for everything. It is the *preparation* of the whole world, and everything is highly *improved* with it – in my opinion: they chose. As it is written: "And you shall season all your grain offerings with salt" (*Lev.* 2,13).[310] If you pay attention to it, then you will discover its greatness. Salt [is found] in all *vegetable* ashes, [in the] vegetable stone, in the calx of the mineral stones, in the substance of the stones, in salts and in everything that is found in the whole world. This is hidden *for its swlywy'w*.[311]

309 In transliteration from the Arabic زرنيخ .

310 The introduction of a Biblical quotation in this passage is a very rare intervention of the Jewish translator/copyist, who customarily adds operative comments.

311 The whole phrase is in transliteration from Italian, but while its first part is easily deciphered as "per la sua" ("for its"), the second part remains obscure.

fol. 21r

(נ״ל כי לזה המאמר רומז מלח הדם . או אומר ורומז על הנשדרא שצריך להרכיב עם מלח אחר שלא יברח ויעלה). ומי שזה ידע כבר ראה סוד סתר המלח. וזהו מלח ספון החכמים שהמלח מלבין הגופו״ וינקם ויתירם ויכבוש הרוחות ויעצרם ויקפיאם. ויצילם מן האש שלא ישרפו. ושים דעתך על המלח. ואל תדקדק אלא בו. הלא תראה כי כל החכמים ישובו לסוף מלאכתם במלח[312] טינומינטו איסו הרבה ואינם רוצים טרטטי אחר אם לא המלח המורכב. (נ״ל שהוא הגוף מקולס כי מאחר שהוא סיד תוכל לעשות ממנו מלח ואותו המלח לא יברח מן האש ויציל הרוחות מן האש שלא ישרפו ויכבשם לא זולת זה) כי מאותו יבא המלח ארמוניאקו הגדול ולא יברח מן האש. וארמס אמר[313] סבון של חכמים ותנקה מה שתחמוד.

ויהיה לך הטוב הגדול הזה מאותו שיש לך וצריך והוא הרבה חזק לסכל ולפתי וקל הרבה למשכיל. ואמר המחבר[314] המלאכה הזאת.

גפרית אש. מגניסיאה עפר. כוכב מים. שדנא אויר.

או חלוף שדנא עפר. מגניסיאה אויר.

כן לא היה מחפוש המלאכה הזאת[a] אם לא כי לא ידע הרכבת מה שגנזו החכמים וגם בספרים מחקוהו ואותו לא למדוהו בגלוי ואני נשבע בשם יודע הכל כי מי שלא ידע המלח הגנוז שום דבר לא יצליח ולא יעלה בידו מן המלאכה אותה שרוצה להגיעה. כי לא ידע הנסתר שלה והסוד. והוא כמי שרוצה לירות החץ בלא מיתר.

[§ 4] **עניין** המלאכה הזאת הוא שתקח ממנו כרצונך וכתוש היטב ושים בקדרה וכסה במכסה ושים בתנור שאופין בו הלחם. והנח לילה או יום. ותתירנו במים מתוק זך בכפלים וסנן. ואחר תקפיא ויתלבן כשלג. אז שים למשמרת ואל תעשה עמו אם לא כשתתירנו ואם תעשה בעניין אחר לא יהיה לך תועלת אם לא יהיה מותר לפי

---

[a] בזאת

---

[312] Blank space in the manuscript, possibly due to a corruption of the *Vorlage*.

[313] Blank space in the manuscript, possibly due to a corruption of the *Vorlage*.

[314] Blank space in the manuscript, possibly due to a corruption of the *Vorlage*.

fol. 21r

(In my opinion: because on this [topic] the discourse hints at the salt of blood, or talks about and hints at *sal ammoniac*[315] which one needs to combine with another salt that does not flee nor sublime). Whoever knows this has already seen a secret that salt hides, which is that this salt is the *soap* of the sages, that salt whitens the bodies, cleans them, melts them, holds back the spirits, slows them down, congeals them and saves them from fire, so that they do not burn. Pay attention to salt and do not focus on anything else. Do you not see that all the sages have returned to it at the end of their operations?[316] *Obtaining it* [is done in] many [ways]. We do not want to *deal with* anything other than composite salt (In my opinion: it is the calcined body since from another body, which is calx, salt can be made, and that salt does not flee from fire and saves the spirits from fire so that they do not burn and they cannot be controlled without it), since the great sal *ammoniac* comes from this and it does not flee from fire. Hermes said:[317] "Soap of the sages." Clean what you desire, and this great good will derive for you from what you have and can [do]. This is very hard for stupid people and amateurs [to grasp], but very easy for the intelligent person. The author said:[318] [in] this operation. Sulphur [is] fire; *magnesia* [is] earth; mercury is water; amethyst is air. Or alternatively: amethyst is earth and *magnesia* is air. **Therefore** no one sought this operation if not he who did not know the path that the sages concealed, that they even deleted from the books, and that they did not teach explicitly. I swear to God, who knows everything, that he who does not know the hidden salt will not succeed in anything, and the operation that he desires to obtain will be out of his reach, since he ignores what is hidden and the secret. He is like someone who wants to shoot a bow without a bowstring.

[§ 4. Preparation of sal ammoniac]

Topic. This operation is that you take as much as you want of it, crush it well, put it in a pot and cover it with a lid. Put [the pot] in a bread oven. Let it rest of a night or a day. Dissolve it in double its amount of fresh and pure water and filter [it]. Then congeal it: it will be white as snow. Now put it in a safe place and do not use it unless you have dissolved it. If you act in a different way, you will not obtain anything useful unless it is dissolved, since

[315] In transliteration from the Arabic نشادر in all occurrences in this section.

[316] The Hebrew text leaves here a blank space, possibly a *lacuna* due to the corruption of its *Vorlage*.

[317] The Hebrew text leaves here a blank space, possibly a *lacuna* due to the corruption of its *Vorlage*.

[318] The Hebrew text leaves here a blank space, possibly a *lacuna* due to the corruption of its *Vorlage*.

fol. 21v

שהוא אז חם ולח. [§ 5] רייזוני של סל אלקלי והוא סל צילא. **דע** כי מלח אלקלי הוא המובחר מכל המלחים ויותר נאה מכולם. והוא יותר ראוי לערב עם מלח משי. וי״א מסי כי הוא יקפיאהו ויעשהו עומד קיים. ועל כן שמו להם אלו השמות. צואי אדון באבנים ומן הדברים הקיימים כי ישוב אותם קיימים עומדים. ועל כן הוא חזק מכל המלחים. לפי שיעשה מן הדברים בייטיבלי והם העשב״ ובייטבילי יש ב״ כחות והוא כח מנירה. וכח בייטבלי. ואותו שיעשה מבעלי חיים יגבר עליו כח שלישי. פי״ משתן או מעצמות ודם ושער. כי יש לו כח מנירא ובייטבלי ואנימליטטי כי אלו הכחות נתוספו להם עם כח אנימלי. ושים דעתך במלח אלקלי ובאבנים הלבנים הנקראים גוייזי לבנים. כי המלח הכבד קלריפיקא ליגוייזי ויחזירם לזכוכית לבנה[a]. והמלחים האחרים אין להם זה הסגולה ובחיי כי נסיתי המלח אלקלי ושמתי על האש ובמהרה יותך ויותר מכל המלחים שראיתי כי המלח אלקלי יש לו איקונייציאוני לבנה בלא שרפה. נ״ל כי זהו סוד גדול כי אם תדקדק לשים מלח מאודם בגופות שים זה שלא ישרוף. אם כן זה המלח די לנו ונניח האחרים. וטבעו הוא חם ולח כי הוא אוגיינטו אותו שהקפיאו יובש הארץ. ואמרו אחרים כי טבעו חם ויבש. והוא קיים עומד אצל האש וינצח כל דבר מעופף אותו שהוא מן המעקן ויעצור הטלק. ויכבשנו באש. ויתיר כל הגופות. אם כן די לך זה המלח. ותהיה די לך אם תניח האחרים. כמו שהזכוכית אשר הוא עמו כי עמו יקלסו ליגוייזי הן אותם האבנים ועמו תתיכו ויתירו וכן בלא ספק יהיה מלאכתו בגופות ובעקקיר אם כן יש לך העקקיר שבו הוא החכמה אשר החכמים הראשונים חשבו. וזה המחשבה היה מיד ביליוסמינטי נסתר ומלאכה גדולה. עתה אני נשבע לך בשם כי

---

[a] לבן

fol. 21v

it is now hot and humid.

[§ 5. On alkali salt]
Method for *alkali*[319] *salt*, which is the salt *ṣyl'*. **Know** that *alkali* salt is the most precious of all salts and the most beautiful among all of them. It is the most suitable for mixing with *dough* salt – according to some: *dough* –[320] since it congeals it and makes it stable and durable. For this reason, it is called by them with these names, *which are* the master of the stones and of the stable things, since it makes them durable and stable. For this reason, it is the strongest of all salts, since it is made from *vegetable* things, that is herbs. And *vegetable* things are provided with two faculties: the *mineral* faculty and the *vegetable* faculty. Regarding what is made from living beings – for example, from urine, bones, blood, and hair – a third faculty prevails in them, since it has the *mineral* and *vegetable* faculty and [the faculty of] *animality,* since those faculties are joined by the *animal* faculty. Pay attention to *alkali* salt and to the white stones called white *gwyyzy*,[321] since salt increases *the clarity of the gwyyzy* and reverts them to white glass. The other salts do not have this quality. [I swear] on my life that I tried *alkali* salt and I put it on fire and it dissolved quickly, more quickly than any other salt that I have seen, because *alkali* salt has *a* white *condition* without combustion – in my opinion: because this is a great secret, since if you pay attention in putting reddened salt on the bodies, you [should] put this [kind of salt] that does not burn. For this reason, among the salts this one is enough for us and we [can] leave out the others. Its nature is hot and humid, because it is an *oily substance* that the dryness of the earth has congealed. Others said that its nature is hot and dry. It is stable and durable [when exposed] to fire and every volatile thing prevails on it since it is from *hm'qn*. It helps talc and protects it from fire. It dissolves all the bodies. For this reason, it should be enough for you, and it will be enough, if you leave out the others, like [it happens] with the glass that is with it, since with it those stones [known as] *gwyyzy* are calcined. Melt them and dissolve them without hesitation. This is its operation and its remedy on the bodies, and so you will have the remedy in which the wisdom devised by the ancients is found. And this device was *wonderfully* hidden and it is a great operation. Now I swear to you by God

319 In transliteration from the Arabic القلي in all occurrences in this section.

320 The explanatory note here provides an alternative spelling for the transliteration of the Italian (or Latin) word *masse*, which is spelled as משי in the main text and as מסי in addition.

321 This word may be derived from the transliteration of the Arabic *ǧandal*, meaning "flint" or "river stone." The Latin versions have: *lapis fluminis* (Steele ed. § 7) and *ginges* (Ruska ed. § 78).

fol. 22r

כבר גליתי לך הסתר הגדול וידוע הוא [§ 6] לסואה אופירציאוני. והוא שתקח מאלומי דאלסור חלק וכתוש היטב ושים במזוכך. ושים לכל ליטרא ממנו ו״ ליטרא מים זכים מתוקים ובשל באש עד ישוב לחומש. והנח עד שירדו השמרים. וסנן. והשלך השמרים. אחר שים במזוכך והרתח לאט לאט עד שיקפא. ויזך כקריסתלו ויהיה מלח חשוב ותשמח עמו.

[§ 7] מלאכת טנכר שיעשה מן המלח דאלסור שאמרנו.

קח המלח החשוב כרצונך ותתירנו בכפלים כמוהו מים של סיד אחר שתסננו ימים. אחר קח ממלח הלחם כמו מזה המלח. ותתירנו בכפלים מים קר זך. ושים בה ב״ אוקייו״ דבש דבורים ויהיה מותר. אחר ערב הב״ מימות ובשל באש רפה עד שיקפא כעין קריסטלו. ואז תתיר ושתוק. כי זהו מן התועלות הגדולות. ובעת שתרצה להתיר שום גוף קחנו. ואם אותו הגוף שתרצה להתירו יהיה אבק ועפר או לימטורא. או למנא יהיו. טחון אותו היטב ושים במזוכך והדלק תחתיו האש עד שיותרו כמו חלב. אז תסתיר זה המלח החשוב. ותודה להשם. י״ת שנתן לך זה. כי עם זה ירחץ ר״ל פרפררי הגופות ויותרו. ויקפאו. ויכבוש כל בורח וכל מעופף ועמו אינצירא הגופות. ויקשרו הרוח. וזה הסתירו החכמים שמו אותם תחת המכסה. וזהו נשדרא של חכמים. והחומץ שלהם. והמלח שלהם ודע זה.

[§ 8] **שים** א״ חלק לבנה. וג״ חלקים כוכב. ורחץ במים ומלח עד שיחזור לבן צלול זך. ויכנס על זאת המלגמא. נ״ל חלק סיד קליפות ביצה שהיא הרוח הצומח בגוף מכנס עליו המים המורכבים באבן החיונית נ״ל כוכב מותר. וישמע כשעוה הכל בהכנסת המים והצליית עד שיותר על הלשון. וכניס החומץ. נ״ל בהתרה עד

fol. 22r

that I have revealed to you the great secret which is now known.

[§ 6. Preparation of alkali salt]
Its operation is that you take a part of alum *sory*, crush it well and put it in a pot. For every *pound*[322] of it put six pounds of pure and fresh water and cook it on fire until it is reduced to one fifth. Let it rest until the dross settles, filter it and remove the dross. Then put [it] in a pot and boil it slowly until it congeals and purifies like *crystal*. It will be a valuable salt: rejoice with it.

[§ 7. Preparation of borax]
Operation of *borax*[323] that is made from the salt *sory* which I have mentioned. **Take** as much as you want of the valuable salt and dissolve it in twice its quantity of water of calx after you have filtered it for days. Then take as much bread salt as that salt and dissolve it in twice its amount of cold and pure water. Put on it two *ūqiyya*[324] of bees' honey and it will be dissolved. Then mix the two waters and cook them on a gentle fire until it congeals with the appearance of *crystal*. Then dissolve it and be quiet because this is among the very useful things. Use it when you need to melt any body. If the body you wish to melt is dust and ash or *filings* or a *foil*, grind it well, put it in a pot, and light a fire under it until [the body] has melted like milk. Then conceal this valuable salt and thank God, may He be exalted, that has given this to you, since with it the bodies are washed – this means *to prepare* – they are melted, they are congealed, and every fleeing and volatile substance is conquered. *Cerate* the bodies with it. The spirit is tied [with it]. The sages hid it and placed these [things] under secrecy. This is the *sal ammoniac*[325] of the sages, their vinegar and their salt. Know this!

[§ 8. Preparation of an elixir from silver, mercury, salt and eggshells]
Place one part of Moon[326] and three parts of mercury and wash them with water and salt until it returns white, clear, and pure and it enters into this *mixture* – in my opinion: a part of the calx of eggshells, which is the spirit that grows in the body on which the waters composed in the needed stone are united – in my opinion: dissolved mercury. Cerate everything like wax by adding water and by roasting, until it dissolves on the tongue. Let the vinegar penetrate – in my opinion: by dissolving – until

[322] Transliteration from Italian *litro*, but here to be interpreted as *libra*, "pound."
[323] Transliteration from Arabic تنكار .
[324] Transliteration from Arabic أوقية.
[325] Transliteration from Arabic نشادر .
[326] Moon is a common *Deckname* for silver.

fol. 22v

שיותר הכל למים שחורים בקבורה בלחות וכשיותר מה שהוקפא אז נשלם האקציר.[§ 9] רייזוני של נשדרא

דע כי הנשדרא יותר טוב ויותר חשוב מכל המלחים אקוטימנטו כי הוא יתיר הכוכב ויחזירנו למים קורינטו כשתעלהו עמו ותשימנו להתיר במקום לח. והוא אוגיינטו שתקפיאו יובש האש. וטבעו חם ויבש. ודק איפילטריבי מצד לצד. והוא רוח מעופף ועוצר ועוזר אדלסיר. ואם לא בעבורו לא היה נשלם אלקציר. ולא היה נתך ולא היה מתערב אחר עם האחר. ולא היה נכנס אחד באחד. ובו הוא נסתר כל סתר. ובו הוא אנטרנטי אידיסינטי אידינצירטורי מכל הגופות ומיבש שחרות דלופונמנטא של גופות. ומתיר הגופות והוא אבן פועל נוצח קושטטורי דילקונטדאדיטא. ומכל הרוחות עם הגופות. והוא אבן מוליד. ואחר סוסוטריי אבל הסימן נשאר[a] דלגיירציאוני עד עולם. ובו הוא סוד גדול בעת שתתיר אידיסולבירי מלח הלחם. (נ״ל ש״ם אם תתיר נשדרא ומלח הלחם וערב ותקפיא לא יברח. נ״ל וכן עשה עם מלח אלקלי תתיר כל אלו הג״ מלחים ותקפיאם ולא יברח נשדרא) (בינות המלחים[b] תרכיב מלח אלקלי. ומלח הסיד. ומלח ארמוניאקו בשוב. ותתירם בלחות ותסנן היטב ותקפיא ותתיר ז״פ ושמור. וזהו נשדרא הגדול להלבין. ואם תרצה להאדים הוסף בו זאג מאודם כמוהו ותסנן. בראש הס״ משים מלח האפר ג״כ) ותערב ותקפיא עד שיהיה קשה. וזך כברד וכשוהם וככפור. אז יהיה נקשר בשוה ומעורב יחד קיים בכל עת. איקונפיינטא אירטיני אוינוקיללטרו ויהיו קיימים עומדים באש ולא יעשנו ולא ישרפו. ובעת שככה יהיה מקובצים ליבררי כל גוף שתרצה

---

[a] ושאר [b] במלחים

fol. 22v

all melts in dark water by burying it in humidity. When what was congealed is dissolved, then the *elixir*[327] will be complete.

[§ 9. On sal ammoniac]
Method for *sal ammoniac.*[328] **Know** that *sal ammoniac* is better and nobler than all the salts for its *preparation,* since it melts mercury and reverts it into *running* water when you sublime it [with sal ammoniac] and you put it to dissolve in a humid place. It is an *oily substance* that the dryness of fire has congealed. Its nature is hot and dry, thin and *filterable* from one side to the other side. It is a volatile spirit that retains and assists the *elixir.* Were it not for it, the *elixir* could not be completed, no one could melt or mix one [body] with another and no one could make one [body] penetrate another. Every secret is hidden in it. In it there is the *entering and the exiting and* [it is] *the cerator* of all the bodies and what dries the blackness *from the foundation* of the bodies. It melts the bodies, it is the active and winning stone and *what pulls together opposing things* and all the spirits with the bodies. It is the stone that *generates* and then *subtracts itself,* but its sign remains from *the generation* until forever. In it there is a great secret in the moment when you can dissolve *and dissolve* bread salt (In my opinion: this passage explains that if you dissolve *sal ammoniac* and bread salt and mix and congeal them, it will not flee. In my opinion: do like this with *alkali*[329] salt: dissolve all those three salts, congeal them and the *sal ammoniac* will not flee). (Among the salts compose *alkali* salt, salt of calx and *sal ammoniac* again. Dissolve them in humidity, filter them well, congeal and dissolve – this is a commentary – and save them. This is the great *sal ammoniac* for whitening. If you wish to redden, add to it the same quantity of reddened *vitriol,*[330] filter and place salt of ashes on top of what is filtered. This is also a method). Mix and congeal until it becomes hard and pure like ice, like onyx and like frost. Then they will be uniformly united and mixed in one durable thing forever, and *the components will hold one another* and they will be stable and durable in the fire, they will not smoke nor burn. When they are composed in this way, *they will liberate* every body that you desire

[327] In transliteration from the Arabic إكسير in all the occurrences in this section.
[328] In transliteration from the Arabic نشادر in all the occurrences in this section.
[329] In transliteration from the Arabic القلي in the text in all the occurrences in this section.
[330] In transliteration from the Arabic زاج in the text.

fol. 23r

להתיר. כי הרפואה הזאת יתיר אידיסולבי אותם בע״ה איקומו פלו יתיר גם הברזל היטב. ודע כי לכל דבר צריך המלח קומוני משבח אותו וטוב ידבר ממנו. ושורש הנשדרא הוא מן הצואות ומן לחות האנימלי. אימסימינטי. שי קינגריאה באותו המקום שיעשה האש כמרחץ. ובתחתית אינפונדו הסיר של נחושת. ובאותם סמיאטי פרי ליקוולי לו פוקו סי מוטא תחת הסיר. ואחר ישלימו לסואה אופירא עם העלאה. ועל כן נעשה אדון גדול על כל המלחים כי במלחים האחרים אין רוח כמו לזה והוא אמוטזיאוני שלישית.

[§ 10] ואיכות מלאכתו הוא שתתירנו במזוכך עם המים או בכלי זכוכית. או תשים תחתיו ג׳׳ עלי אפיו או בליטי. ויותר טוב שתעלהו אם תרצה לעשות עמו פירוש לעלות מעל בשני הבדילין או בנחושת או בברזל. ואם תעשה עמו בזהב ובכסף ותעלהו ז״פ. אז יהיה טוב ביותר. [§ 11] איכות מלאכתו הוא שתקח מן הנשדרא חלק וכתוש בכמוהו מלח ותעלהו באלוטילו צואי אֵינְגָלְיאַה רוֹסַדוֹ. ויעלה אותו המלח כמו ג״פ או ז״פ וחדש המלח בכל העלאה ותמצאנו כמו שלג בכיסוי האלוטילו בע״ה.

(נ״ל ביסוד האש אומר כי כחו בנשדרא וכח הנשדרא כאש אך לא יאדים כל כך כמו האש)

המאמר בנפשות וברוחות של מנירא ובזרניך.

[§ 12] **הזרניך** הוא נפש והוא מינירלי טבעו הוא חממות ולחות. ומשני דמיונות הוא. כרכומי ואדום והזרניך שנסיתי הוא כרכומי פולייטו. וזה הזרניך יותר חזק מכל האחרים. ויותר שורף כל הגופות. והזרניך נ״ל האדום הוא פרסימיליינצא לגפרית במלאכות רבות. ובדקות הגפרית. ובמהירות ההתרה. ובלא סבל האש והכרכומי

fol. 23r

to melt, since this medicine melts *and dissolves* them, with the help of God, and *moreover* it also melts iron well. Know that *common* salt is needed for everything: it is praised and spoken well of. The root of *sal ammoniac*[331] is from manure, from *animal* humidity, and *maximally it congregates* in that place where fire is lit, like a bath, at the bottom *at the bottom*[332] of the copper pot and in those [places] *through which fire is lit* under the pot. Then they completed *its operation* with sublimation. For this reason, it is the great master over all the salts, since in all the other salts there is no spirit like there is in it and since it has a third *mutation*.

[§ 10. Preparation of sal ammoniac (i)]
The modalities of its preparation are that you dissolve it in a pot with water or in a glazed vessel, or that you put under it three leaves of celery or of beet. It is better that you sublime it, if you want to operate with it – commentary: sublime with the two kinds of tin, with copper or iron. And if you operate with it on gold or silver, then sublime it seven times. This is much better.

[§ 11. Preparation of sal ammoniac (ii)]
The modality of its preparation is that you take one part of *sal ammoniac*, grind it with the same amount of salt and sublime it in an *aludel*, *that is* in a *galea rosado*. Let this salt sublime like [we said] three or seven times, renewing the salt at every sublimation: you will find it like snow in the lid of the *aludel*, with the help of God. (In my opinion: on the element of fire. It is said that its quality is in *sal ammoniac* and the quality of *sal ammoniac* is like fire, even if it does not redden as much as fire)

The discourse on the souls and the spirits of the *minerals* and on *arsenic*.[333]

[§ 12. On Arsenic]
Arsenic is a soul and it is *mineral*. Its nature is hotness and humidity. It comes in two varieties, the yellow one and the red one. The [kind of] *arsenic* that I have tried is the *clean* yellow coloured one. This arsenic is stronger than all the others and it is the most combustible of all bodies. *Arsenic* – in my opinion the red one – is *very similar* to sulphur in many operations, in the fineness of sulphur, in its quickness to dissolve, and for the lack of resistance to fire. The yellow one

[331] In transliteration from the Arabic نشادر in all the occurrences in this section.
[332] This information is duplicated, once in Hebrew and once in transliteration from the Italian "in fondo."
[333] In transliteration from the Arabic زرنيخ in all occurrences in this section.

fol. 23v

הוא סובל האש פחות מן האדום. והכרכומי צובע יותר טוב איליפואי סטרימטטי דאיסו מיליורא. ומסגולותיו יבא שילבין אחר. אבל לא עצמו אך כשיעלה. ישחיר כשאינו סולימטו. וישרוף כשהוא חי ומה שירצו מן הזרניך הוא שיתלבן ויקחו ממנו השומן שלו. וגם נקח ממנו השריפה שלו. ואמר ייבר בן חיין כי השם יתן לנו עוזר אליספולייזיאוני אמר באיזה. בעת שהזרניך תבשל בשמן שקדים מרים. כל שמן אחר יהיה טוב הרבה. ואמרו אחרים כי החומץ מיליורא איסו דמילייורציאוני אינקורוטיבלי. אלו הב׳׳ זרניכים הם סאינציא. והחומץ בטבע מלאכתו. ואותו החומץ בחיי ראשי שמראה בזרניכים מלאכה נפלאה הבן זה. ואמרו אחרים כי הזרניך הוא אבן העם בזול ונבזה ומושלך בשווקים ובטיט ובמרחצאות אותו שהוא שורש הסיד כשתסתירו כשיגלחו בו שער[a] הנשים. איטיניסי דיאיסירי כסף חי ויתערב בו בעת שיהיה עמו אחיו. אמ׳׳ ליאל נ״ל בלטב. ויט כי בכסף החי יש סוד אחר שלא יהיה בשאר מיני המעדן. ובמעדנים יש קצת הגופות שיש לו כשיתחברו עמו. והוא מטריח החכמים ויעשם עצלים. והקונדימינטו שלו הוא בג׳׳ מינים העלאה. ורחיצה ומנוחה [334]תלבן. [§ 13] וידיעת העלאתו הוא שתקח ממנו חלק נ״ל מלח וטחון ותתיר בבודילו אחד כמו שתדע. ואחר פירמינטה מכלי אל כלי שיסונן היטב וחבר אותו בזרניך כתוש עם טחינה וצלייה עד שיכנס בו לגמרי. ושים בקדרה מזוככת וכסה במכסה פה אל פה[335] וטח היטב בטיט החכמה שיש בו שער. ואחר שים על האש י׳׳ שעות בג׳׳ שעות הראשונות עם אש רפה. ובז׳׳ שעות עם אש חזק. אחר כן תקרר. אז אותו שעלה תקבץ עם ספוגייא. ואם אז ייטב בעיניך הלבנות שלו הרי מה טוב. ואם לאו שים אותו על האש נ״ל קודם התסעיד. ובשל נ״ל ותסעיד עד שיהיה כרצונך כן תעשה שלשה

---

[a] השער

---

[334] Blank space in the manuscript, possibly due to a corruption of the *Vorlage*.
[335] Blank space in the manuscript, possibly due to a corruption of the *Vorlage*.

fol. 23v

tolerates fire less than the red one. The yellow one dyes better and *his offspring* is *better.*[336] Among its qualities is that it whitens another [substance], but not by itself, only if it is sublimed. It blackens when it is not *sublimed.* It burns when it is alive. What [the sages] desired from *arsenic*[337] is that it becomes white; they extract its greasy part from it; its flammability is also removed. Ǧābir ibn Ḥayyan said in [the Book of the] *Abstractions*[338] – may God help us: when you cook *arsenic* in bitter almond oil, or in any other kind of oily substance, it is very good. Others said that vinegar *improves it with an incorruptible improvement.* These two *arsenics* are *science,*[339] and vinegar is in the nature of its preparation. I swear on my life that this same vinegar reveals a marvellous operation for the two *arsenics*. Understand this! Others said that arsenic is the stone of the people, cheap, despised and discarded in the markets, in mud and in the baths: [they said] that arsenic is the root of calx, since it is hidden in it when women's hair are shaved with it. *And it could be* quicksilver; it mixes with it since it is a brother to him. *Ly'l* said[340] – in my opinion: *Balinus*[341] – and said that there is another secret in quicksilver that it is not in the rest of the minerals. Among minerals, there are some bodies that have it when they combine with it. [It] hinders the sages and makes them tired. Its *preparation* is of three kinds: sublimation, washing and rest. [...] Whitens [...].[342]

[§ 13. Sublimation of arsenic]

The knowledge of its sublimation is that you take a part of it – in my opinion: salt, – grind it and dissolve it in a piece of *gut* as you know. Then *ferment it*[343] from a vessel into a vessel so that it is well filtered. Combine it with crushed arsenic by grinding and cooking until it penetrates into it completely. Put it into a glazed pot and cover it with a lid mouth to mouth. Close it well with clay of the sages in which there is hair, then place it on fire for ten hours: for the first three hours on a gentle fire and for seven hours on a strong fire. Let it cool, then collect what has sublimed with a *sponge*. If its whiteness looks good to you, this is good. If not, put it on the fire – in my opinion: prior to the sublimation. Cook – in my opinion: and sublime – until it is as you wish. Do this three

[336] The Italian word here transliterated can be read as "postremitate" or "posteritate" (Engl. "posterity").

[337] In transliteration from the Arabic زرنيخ in the text.

[338] On the citations of Ǧābir and his works, see *Introduction*, § 4.3 and § 5.7.

[339] The word here in transliteration is the Italian rendering of the Latin "scientia."

[340] The identification of this authority is problematic: the reading of this name as Lull (or rather Pseudo-Lull) appears too far-fetched to be proposed.

[341] The identification of this alchemical authority is as puzzling to the contemporary reader as it was to the extensor of the Hebrew note, who proposes—if our reading is correct—the name of *Balīnūs*, the Arabic name of Apollonius of Tyana to whom natural philosophical and alchemical writings were attributed.

[342] The text of the manuscript presents extensive *lacunae* here, possibly caused by damage in its *Vorlage*, that make the reading of the passage very problematic.

[343] The operation of filtration described here leads me to think that the verb "to ferment" here given in transliteration from the Italian "fermenta" is due to an error in the translation (or in its *Vorlage*).

fol. 24r

פעמים לא יותר ובכל פעם תחליף המלח המותר ואז יהיה כמו הכנפורא וכל סתר החכמה וידיעת אם[a] נעשה בטוב ואם לאו הוא שתקח למנא *מ״ה ס״ג י״ח* למנא חמה של כסף ושים מן האבק של זרניך עליה ואם יהיה טוב לא ישחיר הכסף. [§ 14] וידיעת השבח שלו שתשקנו מלח מותר כמו שאמרנו למעלה ותצלנו בתנור הלחם בקלות עד שישתה מן המלח המותר כמוהו. ואחר תרחץ וסננהו עַל מן המלח שלו במזוכך עד שיצא המים מתוק. אחר תייבשהו בכלי א״ ואנצירא עם מלח אלקלי מותר כמוהו עד שתראה שיתלבן כחלב. אחר שים על האש עד שיותר *נ״ל שיותך* ואחר תוציאנו ותשמרנו.

[§ 15] **שער** יותר טוב. קח זרניך כרכומי פולייטו א״ ליטרא וממלח משי ס״א מסי כמוהו ותתירנו בחומץ חזק חדש כפלים כמוהו. ותשקנו לזרניך מעט עם טחינה וייבשות עד שישתהו כלו וצלהו ותרחצהו עם מים זכים עד ימתיקו המים וכן עשה עד שתראהו לבן כשלג. ואחר תשרוף עמו כל גוף שתרצה. והוא הרבה טוב אין למעלה ממנו. [§ 16] וידיעת עשייתו הוא שתקח זרניך חלק וטחון וכתוש היטב וקח מן המלח קודם וצלה עד שיפסיד השנטרידורי שלו ותתיר במים וסנן ותקפיא. ותתיר בחומץ חד כפלים כמוהו ואינצירא ממנו אותו הזרניך והוא שתשחקנו בשמש חם למעט מעט טחון והשקות עד שישתהו כלו ויהיה לבן. וצלהו ורחוץ עד ימתיק ולוש אותו עם חמאה[344] כמוהו ותשמרהו אחר האונציאוני במזוכך וטח אותו כמו שאמרנו פעמים ושים בתנור חם ושים המזוכך עם האפר עם הגחלים כמו שיעשה[345] בלחם תחת האפר וזה לילה א״ ולבוקר תוציאנו וטחון כן תעשה עד שיתלבן ככסף. אחר אז תעשה עמו ותראה מה[346] בע״ה שים א״ דרמי ממנו על ק״ דרמי מנחושת[347] או

---

[a] עם

---

[344] Blank space in the manuscript, possibly due to a corruption of the *Vorlage*.

[345] The copyist repeats in superscript the letter ע that was unclear in his first realization in the text.

[346] Blank space in the manuscript, possibly due to a corruption of the *Vorlage*.

[347] Blank space in the manuscript, possibly due to a corruption of the *Vorlage*.

fol. 24r

times and not more; and every time replace the dissolved salt: then it will be like *camphor.* All is a secret of wisdom. The knowledge of whether one has operated well or not is that you take *a hot sheet* *45, 63, 18*[348] *sheet*[349] of silver and put some dust of *arsenic*[350] over it. If it is good, the silver will not become dark.

[§ 14. Improvement of arsenic (i)]
The knowledge of its improvement is that you imbibe it with dissolved salt, as we mentioned above. Roast it slightly in a bread oven until it drinks the same quantity of dissolved salt. Then wash it and filter it from its salt in a pot until fresh water comes out. Then let it dry in a vessel for one [day] and *cerate* it in the same amount of dissolved *alkali*[351] salt until you see that it has become white as milk. Then place it on the fire until it dissolves *in my opinion: it melts*, then take it out and save it.

[§ 15. Improvement of arsenic (ii)]
A better **chapter.**[352] Take one *pound* of yellow *clean arsenic*[353] and as much *dough* salt – another interpretation: *dough*[354] – and dissolve it in twice their quantity of strong new vinegar. Water the *arsenic* slowly with it by crushing and drying until it drinks in all of it. Roast it and wash it with pure water until the water has made it sweet. Do this until you see it white as snow. Then burn with it every body you wish. This is very good: there is nothing superior to it.

[§ 16. Making silver with arsenic (i)]
The knowledge of its operation is that you take a part of *arsenic,* crush and pulverise it finely, and take some of the previous[ly mentioned] salt. Roast it until it loses *its shrieking,* dissolve it in water, filter it and congeal it. Dissolve it in twice its quantity of sharp vinegar and *cerate* that *arsenic* with it; that is, you grind it slowly in the hot sun, pulverising and watering it, until it drinks it all and it is white. Roast it and wash it until it becomes sweet, and knead it with the same amount of butter. Then save it after the *anointment* in a pot and cover it as we have already mentioned few times. Put it in a hot oven and place the pot with ashes and coal, as it is done with bread, under ashes. This is [for] the first night. In the morning, take it out and crush it. Do this until it becomes white like silver. Then do like this and you will see what [...],[355] with the help of God. Put one *dirham* on one hundred *dirham* of copper [...][356] or

[348] The interpretation of this marginal note, containing Hebrew numerals, eludes an easy interpretation.
[349] The Italian word *lamina* in transliteration is repeated twice.
[350] In transliteration from the Arabic زرنيخ in all occurrences in this section.
[351] In transliteration from the Arabic القلي in the text.
[352] The term שער meaning "gate" and here indicating a new "chapter" mirrors the Arabic usage of the term باب, literally "door."
[353] Here a combination of Arabic and Italian is employed, with "arsenic" in transliteration from the Arabic زرنيخ and "clean" in transliteration from the Italian "pulito."
[354] The short note is limited to the substitution of the letter ש with the letter ס, but does not provide a real synonym nor an explanation.
[355] The text of the manuscript presents a *lacuna* here, possibly caused by damage in its *Vorlage.*
[356] The text of the manuscript presents a *lacuna* here, possibly caused by damage in its *Vorlage.*

fol. 24v

מברזל או מב״ בדילין ותנפח נ״ל עד שיותך ויהיה כסף א״ סמינטו בע״ה [§ 17] **שער** אחר טוב ממנו. קח זרניך כרכומי א״ ליטרא וכמוהו מלח הלחם. וכתוש יחד הדק טוב ושים במזוכך מטויין וטח בטיט החכמה ושים עליו עד שיכסהו ד״ אצבעות סיירו חמוץ. ובשל באש רפה שיבש וכשיבש הסיירו תוסיף ממנו עוד אבל החום של אש רפה עשה עד שיתעבה שם א״ לילה וכשתיבש שים עליו כמוהו חמאה עם הזרניך ששמת אוה א״ ליטרא ובשל באש לאט לאט עד שיותך כחלב. וקח א״ ליטרא מאלונקי הוא בדיל. נ״ל או קח עופרת. ותתיכנו ותטעמנו בכסף חי כפלים ורחץ[357] השחרות עם הזרניך הנז״ אחר שים הכל במזוכך ויהיה צווארה צר וסתום פיה היטב וטח היטב בטיט החכמה ושים על האש ועליה הדלק הגחלים והנח לילה אחד ותקרר ותוציאנו ושים א״ דרמי על א״ אוקייא מנחושת או ציפרו ויהיה כסף בע״ה.[358] אלומי כי הוא אמת בלא ספק אם תדקדק

[§ 18] המאמר בגפרית וטבעו הוא חם ולח כטבע הזרניך[a]. ותיקונו כמו תיקון הזרניך כמו שאמרו החכמים במלאכה הזאת בספריהם כן אני לא עשיתי זה. אבל הגפרית נסיתי עם הכוכב. וסולימא עם אותו עשיתי צנברו טוב ונסיתיו כנסיון הנחושת והיה נחושת שרוף יפה ומופלא. ומה שרוצים ממנו הוא שיקחו השריפה שלו ולקיחת השומן שלו. ויהיה לבן כמו שהוא בזרניך בשוה. והטוב שבגופרית הוא האדום. ואמר ייבר בן חיין כסף השעה. כי הזרניך האדום יעשה בצביעת האדום ב*נ״ל כ*גפרית. והלבן ג״ך בליבון כזרניך ואני לאו נסיתיו. [§ 19] ותיקונו הוא שתקח ממנו א״ ליטרא וכתוש אותו עם אלומי יימיני כמוהו. ושים עליו השתן ימים ד״ שיכסהו

---

[a] כזרניך

---

[357] Blank space in the manuscript, possibly due to a corruption of the *Vorlage*.
[358] Blank space in the manuscript, possibly due to a corruption of the *Vorlage*.

fol. 24v

of iron or of the two tins, blow – in my opinion: until it melts – and it will be silver, *and smynṭw*,[359] with the help of God.

[§ 17. Making silver with arsenic (ii)]
Another **chapter** better than the previous. Take a *pound* of yellow *arsenic*[360] and the same quantity of bread salt. Pulverize them together very finely, put them in a clay pot, and cover it with the clay of wisdom. Put four inches of *whey* of ferment over it until it is covered. Cook on a gentle fire [until] it dries and, when the *whey* is dried, add more of it, but keep the heat of the fire gentle until they are mixed there for one night. When it dries out, put on it the same amount of butter with the *arsenic,* of which you will have taken one *pound*. Cook slowly on the fire until it dissolves like milk. Take one *pound* of *tin*,[361] which is tin – in my opinion: otherwise take lead. Melt it and feed it with twice its quantity of quicksilver. Wash [...][362] the darkness with the mentioned *arsenic*. Then put everything in a pot whose neck is narrow, seal its opening well, and cover it well with the clay of wisdom. Place [it] on the fire and light some coals over it. Let it rest for one night, then let it cool and take it out. Put one *dirham* on one *ūqiyya*[363] of copper or *copper:* it will be silver, with the help of God. [...][364] Alum, since it is true, without doubt, if you pay attention.

[§ 18. On sulphur]
Discourse on sulphur. Its nature is hot and humid like the nature of *arsenic*. Its improvement is like the improvement of *arsenic*, like the sages said in [their] books regarding this preparation. I have not tried this though. I have tried instead sulphur with mercury, and, by *subliming* with this, I have made good *cinnabar*. I have tried to make a trial with copper and the copper became burnt, good and marvellous. What they[365] desire is to remove its combustibility and extract its oiliness: it will become white just like *arsenic*. The best among [the kinds] of sulphurs is the red one. Ǧābir ibn Ḥayyān said: "Make silver because red *arsenic* works for the red dye in – in my opinion: like – sulphur. In the same way, the white for whitening is like *arsenic*. I have not tried it.

[§ 19. Preparation of sulphur]
Its improvement is that you take one *pound* of it and crush it the same quantity of Yemenite *alum*. Put urine on top of it for four days so that it covers it.

[359] The Italian words in transliteration here could be read as "e cemento" ("and cement"); this expression could possibly derive from the Latin *e cemento*, i.e. "silver from the cementation", which is very pure silver.
[360] In transliteration from the Arabic زرنيخ in all occurrences in this section.
[361] In transliteration from Arabic آنك in the text.
[362] The text of the manuscript presents a *lacuna* here, possibly caused by damage in its *Vorlage*
[363] In transliteration from Arabic أوقية in the text.
[364] The text of the manuscript presents a *lacuna* here, possibly caused by damage in its *Vorlage*
[365] "The alchemists" or "the philosophers" is the subject of this sentence.

fol. 25r

ואחר תבשלהו באש רפה עד שיחסר כל השומן וכן תעשה עד שייטב בעיניך דמות לבנותו. ואם לא נתלבן היטב שים עליו סיירו חמוץ חם. ובשל עד שיתלבן. ושים אותו במלאכתך בע״ה.

[§ 20] **המאמר** בכוכב . דע כי הכוכב טבעו קר ולח. והשם ברא ממנו כל המנירי אם כן הוא נשימנטו מאותן. והוא אוירי בורח מן האש. ועם כל זה כשיעמוד זמן עם האש יעשה עשיות גדולות מופלאות וגבוהות. והוא לבדו הוא הרוח חי. ובעולם אין כמוהו ושיוכל לעשות כמו שיוכל לעשות הוא. והוא יכנס אל כל גוף. ר״ל יפלש ויקום[366] גביה הגופות אם כן בעת שיתערב עם אחר מכל הגופות יחייהו ויאיר אותו ויעתיקהו. סטטו אדלטרו סטטו. ומדמות על דמות כשיהיה מעורב עמו. והוא יהיה שאור מאותו הגוף (נ״ל כי לזה רומז כי הוא שאור לכל השאורים) ואז יהיה כולו איקציר. ללבן ולאדום. והוא מים נצחי. והוא מי החיים. והוא חלב הבתולה. והוא עשב מנקה ורוחץ. והוא מקור החיים. כי מי שישתה ממנו יחיה ולא ימות לעולם. והוא מקבל המראים ורפואת הדמומת *נ״ל הדמות* והוא יעשה הגופות מקבלי המראים. והוא ממית ומחיה בע״ה. והוא מיבש ומלחלח. והוא מקרר ומחמם. והוא יעשה מעשים הפכיים כפי לוסואו קונדימינטו. ובעת שהוא חי יש לו מעשים רבים. וכשיהיה מת יש לו מעשים רבים יותר יפים. וכשיהיה סולימטו יש לו מעשי״ אחרים. וכשיהיה מותר יהיו לו מעשים יותר גדולים. והוא סורפינטו המוליד עצמו כמי אילוסודיאינטי סיבי מידימי[367] עצמו כמי וליום א״ יחיל וילד . והוא ממית כל החיים אלמליאה עם הסם שלו. והאש יבלה ויהרוג כל דבר בזמן מועט או זמן ארוך. ועכ״ז לא יוכל לנצח הכסף חי. כי הוא יברח ממנו. אמנם עשו החכמים בערמה אחת או בתחבולות רבות קואספוטסי האש למעט מעט. ואז מעלה מעלה מדרגה מדרגה. ילחם עם האש ויטעימוהו ג״כ מאש עד שיקוה

[366] Blank space in the manuscript, possibly due to a corruption of the *Vorlage*.
[367] Blank space in the manuscript, possibly due to a corruption of the *Vorlage*.

fol. 25r

Then cook on a gentle fire until all the oiliness is consumed. Do like this until the colour of its whiteness looks good to you. If it does not whiten well, put on it hot whey of ferment and cook it until it whitens. Put this in your operation, with the help of God.

[§ 20. On mercury]
The discourse on mercury. Know that the nature of mercury is cold and humid, that God created all the *minerals* from it, and that for this [reason] it is their *birth*. It is air that flees from the fire, even if, when it stands with fire for some time, it makes great, marvellous and praiseworthy actions. It alone is the living spirit. In the world there is nothing like it that can perform what it can perform. It penetrates all the bodies – this means: it penetrates and remains [...][368] – and brings together the bodies. For this reason, when it is mixed with any of all the bodies, it vivifies it, enlightens it, and makes it change from *one state to another* and from one shape to another. When it is mixed with it, it is the ferment of that body (in my opinion: because in this is concealed the fact that mercury is the ferment for all the ferments). Then it is the complete *elixir*[369] for the white and for the red. It is the eternal water, the water of life, the milk of the virgin, the herb that purifies and washes, the source of life, because he who drinks from it will live and will never die.[370] It can receive different appearances and it is the medicine of the forms *in my opinion: forms*.[371] It makes the bodies receptive to the appearances. It is dead and living, with the help of God. It is the one who dries and the one who humidifies, the one who cools and the one who warms. It performs opposite things depending on *its preparation*. When it is alive, it does many things. When it is dead, it does many things that are even better. When it is *sublimed*, it does other things. When it is dissolved, it does even greater things. It is the *serpent* that generates like mellow water *itself*[372] by itself, like water; in one day it is born and gives birth. It kills all the living *animals* with its poison. Fire causes corruption and kills everything in a short or long time, but, despite all this, it cannot defeat quicksilver, because it flees from it. Nevertheless, the sages, through one expedient or many tricks, made it *withstand* fire little by little, more and more, step by step. It fights with fire, but it is also eaten by it until it weakens

[368] The text of the manuscript presents a *lacuna* here, possibly caused by damage in its *Vorlage*.
[369] In transliteration from the Arabic إكسير in the text.
[370] This phrase bears a striking similarity to a passage of the ritual of the consecration of wine and bread in Christian Catholic rite, thus corroborating the hypothesis of an Italian (or at least Catholic) environment for the production of the Hebrew translation of the *OAS*.
[371] The marginal note here proposes a simple correction of spelling, from דמומות to דמות.
[372] The text of the manuscript presents a *lacuna* here, possibly caused by damage in its *Vorlage*.

fol. 25v

וימתין אותו נ״ל מיסוד האש או גפרית נ״ל פי אלקוינטי אשפיטזיאוני. ואז יהיה ממנו מעשים גפלאים ההעתקה כפי מה שהעתק כן יעתיק כי אז נלקח ממנו שחרותו ויאירו[a] ונלקח הזכות[b] שלו. אם כן אז כמו שנצבע כן יצבע וכמו שנקפא כן יקפיא. ובעת שיותר יהיה מותר ויתיר אחרים. והוא ילבין עצמו כמי[373] כמו שנראה יראה. ולבסוף יאדים. והוא חומץ מר. והוא בין חומץ[c] למר. והוא מים המחבר. והוא חלב. ושתן לח. והוא אוגיינטו סולוטיבו. והוא אבי כל הפלאות. והוא איד האידי. והוא עבד הבורח. והוא נתווכח עם הזהב ונצח אותו. ואמר הזהב התתווכח עמי. אני אדון האבנים ואסבול על האש. ענה הכוכב אמת תאמר. אבל אני ילידתיך וממני תוציא תולדה. וחלק א״ ממני יעשו אלף חלקים ממך ואתה כילי כי שום דבר לא תרצה לתת משלך בערך אלי. ומי שיחברני עם אחי או אם אחותי יחיה וישמח ואהיה די לו לעולם ואפילו אם יחיה אלף אלפי שנים. ואני הוא כל הסוד הנסתר. ובי הוא ר״ל החכמה הנסתרת. ואני אעתיק כל הגופות ללבנה עם זרניך ר״ל כי כן צריך להלבין תחילה. ואחר אעתיקם לשמש עם גפרית. ובי הם ב״ מינים תולדות דיקונדימינטו הא״ מהם מלח ואטרמנטו אסילטויאו. וגם הנשדרא. וגם עם הגופות טעימה. ועם ריח בדיל אולפטו. ועם אולפטו ריח הגפרית.

[§ 21] **והעלאתו** עם המלח והזאג הוא שתקח ממנו חלק וטחון אותו הכוכב עם הזאג בכפלים. ועם המלח

---

[a] ויאירי [b] היזכות [c] חמוץ

---

[373] Blank space in the manuscript, possibly due to a corruption of the *Vorlage*.

fol. 25v

and it is made moderate – in my opinion: by the element of fire or by sulphur – in my opinion: *with a long wait*. Then, many excellent operations of transmutation derive from it, because everything that it causes to transmute is transmuted in that way, taking from its darkness, its brightness and its purity. In the same way, as it is dyed, so it dyes, and as it is congealed, so it congeals. When it is dissolved, it is dissolved and dissolves other [metals]. It whitens by itself like who [...][374] and as it is shown, it appears. Finally, it reddens. It is a bitter vinegar and it is among vinegars for bitterness. It is the water of the combiner; it is milk and humid urine. It is an *oil that dissolves*. It is the father of the marvelous, the misfortune of the misfortunes and the fleeing servant.[375] Mercury had an argument with gold and won. Gold said: "Are you arguing with me? I am the lord of the stones and the one who withstands fire." Mercury answered: "You spoke the truth. But I have generated you and the generation derives from me. One part of me generates one thousand parts of you. You are stingy, because you do not want to give anything that you have, as opposed to me. Whoever mixes me with my brother or my sister lives and rejoices, since I will be enough for him forever, even for a thousand of years.[376] I am the whole hidden secret and it is in me – this means: the hidden wisdom. I transmute all the bodies into the Moon with *arsenic*[377] – this means: since one must whiten in this way at the beginning. Moreover, I transmute them into the Sun[378] with sulphur. In me there are two kinds of accounts for *preparation*: the first of them is [with] salt and *atrament and sublimation* and also [with] *sal ammoniac*;[379] also with the bodies, through nutrition, and with the *smell* of the exhalation of tin, and with the *smell* of the exhalation of sulphur.

[§ 21. Sublimation of mercury (i)]

Its sublimation with salt and *vitriol*[380] is that you take one part of mercury and grind it with twice as much *vitriol* and with

374 The text of the manuscript presents a *lacuna* here, possibly caused by damage in its *Vorlage*.

375 This passage lists some of the most common *Decknamen* for mercury.

376 The brother and the sister in this passage can possibly be identified with gold and silver, whose quantity is believed to be increased when in combination with mercury.

377 In transliteration from the Arabic زرنيخ in the text.

378 Sun is a common *Deckname* for gold.

379 In transliteration from the Arabic نشادر in the text.

380 In transliteration from the Arabic زاج in all occurrences in this section.

fol. 26r

בכפלים או כמוהו והמלח יהיה מותר בבודילו. וטחון עם אלו השנים ויבש עד שימות בהם. אחר שים באלוטילו וראה והזהר שלא יהיה בו לחות. וקודם תצלהו בתנור הלחם הכוכב לילה אחד ובבוקר תוציאנו ותשקנו המלח המותר אחר שים בתנור כן עשה הרבה פעמים ונ״ל מה עד שישחיר הוא. וכל מה שיותר תשים בתנור כך יותר הרבה יקבל דמות אדמימות. אז תשים אותו באלוטילו וסתום הפרק היטב ועשה תחתיו אש לאחר שתקנת אותו על האש שלשה שעות תעשה אש רפה. ואחר תגדיל האש ז״ שעות אם הוא בקייץ. ואם הוא בחורף עשה מבוקר עד ערב. אחר הנח עד שיקרר. ואז פתח האלוטילו ותמצא למעלה מעולה לבן וזך בע״ה.

[§ 22] **והעלאתו** עם הנשדרא הוא שתקח ממנו. ר״ל מן הכוכב ב״ חלקים ושים א״ חלק נשדרא לבן ביניקונבולטו. ושים מן המלח א״ חלק. וטחון היטב עד שימות בו הכוכב. ואחר שים באלוטילו כמו שאמרנו. יעלה אז באלומי לבן לנוסו. וזה א״ מן המלאכות הגדולות וגבוהות ותשמרנו והנח.

[§ 23] **הקפאתו** עם הגפרית הוא שתקח ג״ אוקייו שמן שירג״ אם תמצא ואם לא קח שמן זית. והרתח על האש בכלי זכוכית ואחר כן שים עמו חצי אוקיי״ מגפרית כרכומי כתוש למעט מעט וטחון עם השמן עד שיותר. נסיתיו והותר כלו. אחר תורידנו מעל האש והנח לקרר בו. נ״ל שצריך לעשות זה שיהיה לבן כעופרת. ואז שים עמו א״ אונקיי״ מכוכב. ואחר שים הקדרה על האש הרפה[a] לאט לאט ולמעט מעט הדלק והנח עד שיקפא. כי יקפא הכוכב לאבן אדום. נ״ל לבן אם לא תשים בו יסוד האש. אחר תסתירנו ותשמרנו. וותשמח. ושים ממנו בכל מלאכותיך

[a] רפה

fol. 26r

twice the amount or an equal amount of salt, and the salt has been dissolved in a *sausage skin.* Grind it with these two and dry it until it dies in them. Then put it in an *aludel* and check to make sure that there is no humidity in it. First, roast the mercury in a bread oven for one night and, in the morning, take it out, and water it with the dissolved salt. Then put it in an oven. Do this many times – and in my opinion: as many as it takes until it becomes dark. Every time that it dissolves, put it again many more times in the oven until it receives a red colour. Then put it in an *aludel,* seal the junction well and light a fire under it. After you have placed it on the fire for three hours, make the fire gentle; after that, increase the fire for seven hours, if it is summer, or from the morning to the evening, if it is winter. Then let it rest until it cools. Then open the *aludel* and you will find on top pure white sublimate, with the help of God.

[§ 22. Sublimation of mercury (ii)]
The sublimation with *sal ammoniac*[381] is that you take two parts of it – this means: of mercury – and put [on it] one part of white *well convoluted*[382] *sal ammoniac.* Put one part of salt. Pulverise it well until the mercury dies in it. Then place it in an *aludel* as we already mentioned: it will sublime then in white alum *like wool.* This is one of the great and exalted operations: save it and rest.

[§ 23. Congealment of mercury]
The congealment with sulphur is that you take three *ūqiyya*[383] of sesame oil, if you find [it] or, if not, olive oil. Boil it on fire in a glass vessel and, after this, put with it slowly half an *ūqiyya* of pulverised yellow sulphur, and grind it with the oil until it dissolves. I tried it and I dissolved it all. Then take it out on top of the fire and let it rest to cool in [it] – in my opinion: you should do it like this and it will be as white as lead. Then put with it one *ūqiyya* of mercury, place the pot on a gentle fire, and increase it slowly and little by little. Let it rest until it congeals, since mercury will congeal to a red stone – in my opinion: white. If not, put in it the element of fire. Then conceal it, save it, rejoice and put [some] of it in all your operations.

[381] In transliteration from the Arabic نشادر in all occurrences in this section.
[382] This expression is transliterated from the Italian "bene convoluto"; its meaning is not clear in this context.
[383] In transliteration from Arabic أوقية in all occurrences in this section.

fol. 26v

[§ 24] **הקפאתו** עם ריח הגפרית. וזהו שתקשור הכוכב בבגד מהודק. (בתקוני ייבר מפרש ואומר כי זה הבגד רומז כדור מטיט.) ותלהו בקדרה ושים אלופונו סואו גפרית וקשור בקשר רך הבגד וטח הפרק היטב מלמטה ומלמעלה. והדלק אש תחתיה א״ יום שלם והנח עד שיתקרר. ואחר כן תמצאנו אדום כמו צנברו. אחר תשימנו במלאכותיך האדומים. ועמו תשמח כי בו יסתר סתרים מופלאים גדולים בלא מספר. או אם תרצה להתירו תתירנו עם נשדרא. ואם תרצה הכוכב מוקפא יהיה לך מוקפא. ואם תרצה אותו ללבן תכתשנו עם מלח אלקלי — (עיין וכוין עם ספר ל״ו בסופו ועם ספר התרת ג״ רוחות) ועם צירוסי ותתעכב עליו הטחינה ר״ל שתטחנו ותשקנו ימים ויתלבן. ויהיה כמו אם היה סולימטו שילוטילו. ואם תרצה שישאר הצבע שלו אדום לעולם מאותו המוקפא אדום תכתשהו עם אטרמנטו ועם קרוקו של ברזל ועם מי הנשדרא ר״ל מאודם אז ישאר אדום ואדמותו יעמוד. אחר תסתירנו ותשמרנו.

[§ 25] **הקפאתו** עם ריח האסרוב. זהו שתקח א״ כדור משעוה וטח סביביו מטיט החכמה ובעת שיתייבש עשה נקב א״ וחמם על האש עד שימס השעוה. ושים בו הכוכב. נ״ל רחוץ בג״ אומורי בבשול עד שיתמלא היטב וסתום הנקב היטב מאד ושים בקדרה א״ שיהיה בה עופרת מותך . והדלק האש תחתיו ב״ שעות ויקפא לאבן לבן ותסתירנו ושמור ושים למלאכתך ועמו תשמח.

fol. 26v

[§ 24. Congealment of mercury (ii)]

Its congealment with the exhalation of sulphur. It is that you tie mercury in a tight cloth. (In the *Rectifications* of Ǧābir it is explained and said that this cloth is an allusion to a ball of clay).[384] Hang it in a pot and put sulphur *on its bottom*. Tie the cloth with a loose knot and cover the junction well from below and from above. Light a fire under it for a whole day and let it rest until it cools down: then you will find it red like *cinnabar*. Put it among your operations for [obtaining] red and rejoice with it, because innumerable wonderful and great secrets are hidden in it. Otherwise, if you want to dissolve it, dissolve it with *sal ammoniac*.[385] If you want congealed mercury, it will be congealed for you. If you want it white, crush it with *alkali*[386] salt – (look up and consider the end of the 36th book and the *Book on the Dissolution of the Five Spirits*)[387] – and *cerussa*, repeat the grinding – this means: grind it –, water it for days and it will become white: it will be as if it had been *sublimed in the aludel*. If you want its tincture to remain red forever from this red congealment, pulverise it with *atrament*, with the *crocus* of iron, and with water of *sal ammoniac* – this means: reddened. Then it will remain red and the redness will endure. Then hide it and save it.

[§ 25. Congealment of mercury (iii)]

Its congealment with the exhalation of *lead*.[388] This is that you take a ball of wax and cover it all around with clay of wisdom. When it dries, pierce one hole and warm it on fire until the wax melts [out]. Put mercury into it – in my opinion: wash with the three *humours* by cooking – until it fills up well. Close the hole very well and put [the ball] in one pot in which there is molten lead. Light a fire under it for two hours: it will congeal into a white stone. Hide it, save it, put it among your operations and rejoice with it.

[384] On the quotations of Ǧābir ibn Ḥayyān, see *Introduction*, § 4.3.

[385] In transliteration from the Arabic نشادر in all occurrences in this section.

[386] In transliteration from the Arabic القلي in the text.

[387] The Jabirian treatises referred to in this passage could be the *Kitāb al-tadwīr* ("The Book of the Construction of the Circle"), which is listed as the 36th of the *One Hundred and Twelve Books*, or the *Kitāb al-Layla* ("The Book of the Moon"), which is listed as the 36th of the *Seventy Books*. See Kraus, *Jābir ibn Ḥayyān*, I, 25 and 53. In the list of Jabirian works in Kraus there is no trace of a *Book on the Dissolution of the Five Spirits*.

[388] In transliteration from the Arabic أسرب in the text.

fol. 27r

(בתקוני ייבר ב״ פעמים עשה כן ואחר תעלהו).

[§ 26] **הקפאה** אחרת. זהו שתקח קנה עב לח וטח סביבו בטיט החכמה. ושים בו הכוכב חי עד שיתמלא הקנה אחר כסה הנקב מטיט החכמה. ועשה חפירה כפי דמותה ותחתיה שים א״ רקוע עופרת ועל הרקוע שים הקנה תתיך עופרת אחר ותשפוך על הקנה עד שיתמלא חפירת סביב הקנה עד שכלה יתכסה. והנח עד שיתקרר. אחר טח סביבות העפרת מטיט החכמה והנח ליבש והדלק תחתיו האש ועליו וסביביו עד שיותך העופרת בטיט החכמה וריח העופרת יקפיא הכוכב שבתוך הקנה . אחר הנח לקרר ותוציאנו ויהיה מוקפא החשוב תמצאנו שאין כמוהו. נ״ל אם תעשה פעמים רבות יהיה יותר טוב[389] בכוכב . אם כן תסתירנו. [§ 27] ותטעום בגופות שתתיך כל א״ מהם חוץ מן הברזל כי לא יקבלנו בטעימה. ושים עליו גוף שתרצה להתירו הכוכב שאמר[a] והכוכב יהיה בכפל ותרחצנו בעת שיהיה טעום נ״ל או במים ומלח עד שיצא כל השחרות ממנו. אחר תייבש לשמש ואינצירא אותו נ״ל כל אחת מן המימות. נ״ל מי ברזל בין חומצים. נ״ל עוד מי כוכב מותר או נשדרא מוסער ואלומי מותרים. למדין ריטורנטי לאש קלה. או לשמש כן תעשה כל היום. אחר תתירנו. נ״ל אם תקפיא וכשיותר אנצירא עמו הזרניך מלובן. אחר שים כל אותו הדבר להתיר . וכשיות*ר* תקפיאנו. אחר א״ דרמי ממנו בלבד ילבין ד״ מאות. מנחושת לכסף. מאתים מב״ בדילין ב״ה. ודע כי ההתרה הוא שורש אלו הדברים. ובו הוא כל הסוד אם תתיר אותו כמו שתרצה או בזבל או במרחץ . א״כ תתיר עמו כל הסידים וכל הלימטורי אם רצון השם. אמר ייבר בן חיין כמי שיחון לו השם אמר באמת

---

[a] דאמר

---

[389] Blank space in the manuscript, possibly due to a corruption of the *Vorlage*.

fol. 27r

(In Ǧābir's *Rectifications*:[390] do this twice and then sublime it).

[§ 26. Congealment of mercury (iv)]
Another **congealment**. This is that you take a thick and moist reed and cover it all around with the clay of wisdom.[391] Put into it live mercury until the reed is full; then close the hole with the clay of wisdom. Cut a channel in the shape of the reed, put under it a thin layer of lead and place the reed on top of it. Melt some other lead and pour it on the reed until the channel around it is filled and completely covered. Let it rest until it cools. Then cover around the lead with clay of wisdom and let it rest to dry. Light a fire under it, over it, and around it until the lead melts in the clay of wisdom and the exhalation of lead congeals the mercury that is in the bottom of the reed. Then leave it to cool and extract it: it will be congealed and noble. We have not found anything like it – in my opinion: if you do it more times, it is better. [...][392] in mercury. If so, hide it.

[§ 27. Nutrition with the bodies]
The nutrition with the bodies is that you melt each one of them, apart from iron, which does not accept the nutrition. Put the mercury I mentioned on the body that you wish to melt. Let the mercury be double [the amount of the body]. Wash it when it has been eaten – in my opinion: or with water and salt until it all its blackness comes out of it. Then dry it in the sun and *cerate* it – in my opinion: each of the waters – in my opinion: water of iron among vinegars – in my opinion: also water of melted mercury or sublimed *sal ammoniac*[393] or dissolved *alums and return to put them* on a gentle fire or in the sun. Do this for the whole day and then melt it – in my opinion: if it congeals and when it melts, *cerate* the whitened *arsenic*[394] with it. Then put all of this to dissolve and, when it dissolved, congeal it. Then one *dirham* of it alone whitens four hundred [*dirham*] of copper into silver and two hundred of the two [kinds of] tin, with the help of God. Know that dissolving is the root of these things and in it there is all the secret, if you melt it in manure or in a bath, as you wish. Then dissolve with it all the kinds of calx and *filings*, God willing. Ǧābir ibn Ḥayyān, may God have mercy of him, said: 'Truly

[390] On the quotations of Ǧābir ibn Ḥayyān in the *OAS*, see *Introduction*, § 4.3.
[391] The recipe prescribes the use of a moist reed, which probably means a green reed, not dried out yet.
[392] The text of the manuscript presents a *lacuna* here, possibly caused by damage in its *Vorlage*.
[393] In transliteration from the Arabic نشادر in the text.
[394] In transliteration from the Arabic زرنيخ in the text.

fol. 27v

האסלטציאוני כן הוא של רוחות כמו שהוא ההתרה לגופות. ונזכור אח״כ ההתרה של גופות ב״ה

[§ 28] **המאמר** בגופות. דע כי הגופות מינירליאה הן אלקוונטי[395] אדום עביד[a] שנקפאו כפי מעשה הטבע בזמן ארוך. ואותו שבתחלה נקפא כי הוא כסף חי. והגפרית. והן נשימינטו דמנירא. והם מים ואוגיינטו דארב. אם כן עליהם בישול מחמימות שוה עד שיעבו וממנו יולדו הגופות ואח״כ נעתקו מדרגה מדרגה עד שיעשה זהב וכסף באלפי שנים. ואם דורסורו אלו הגופות במנירי יעשה בהם הטבע עד שיעשה כסף וזהב. אבל השם שם בזה העולם הקטן טבילי שהוא כמו העולם הגדול כל כך חכמה שיוכל לעשות ביום אחד אלו של כסף וזהב. וקראו גפרית דטירה. וקונדימינטו סואו הוא בשלש מינים. ס״א הוא ד״ מינים. הראשון הוא קלצינמינטו עם מלח. והתרה ואח״כ הקפאה ואינציררי. אבל לקלצינמינטו הוא[396] יפה באש חזק עם המלח שיוקח ממנו הגפרית שישחית אותו. וישארו אז נקיים כמו הסיד. והוא[397] אמר ייבר בן חיין כי היו צריכים לקלס אלו הגופות. שינקו החלקים שלהם ושיהיו מוכנים אפילטריזיאוני של לחות בכלי[398] לוקוולי אם כן יהיה ממשלת חלק להתירם ויתחזק היובש שלהם . ואז יהיה יותר או לקבל הלחות ויותר מוכן להתרה. נ״ל ויהיו כמו הזהב והכסף. והוצרכו לקלס ב״ הבדילין והנחושת. והברזל. בעבור הדבר שאמרנו. ושינקו אותם מן הגפריתים שישחיתו אותם ושינקו אותם מן השמר *כ*ים[399] השורפים. ואותו שהוא קצה יותר מכלם. כך יעשה שתוציא מהם לחותן שישחית

---

[a] עביר

---

[395] Blank space in the manuscript, possibly due to a corruption of the *Vorlage*.
[396] Blank space in the manuscript, possibly due to a corruption of the *Vorlage*.
[397] Blank space in the manuscript, possibly due to a corruption of the *Vorlage*.
[398] Blank space in the manuscript, possibly due to a corruption of the *Vorlage*.
[399] The word has been corrected by the scribe, who deleted the ר of השמרים ("dross") and replaced above the line with a כ for שמכים, a term whose sense eludes me in this context.

fol. 27v

sublimation is for spirits what melting is for bodies. After that we will mention, with the help of God, the melting of the bodies.

[§ 28. On the bodies]
The discourse on the bodies. Know that the bodies are *mineral* and that they are *many* [...][400] red, cloudy, that has congealed according to the working of nature in a long time. Those that congealed at the beginning are quicksilver and sulphur and they are *the birth of minerals.* They are mostly water and oily substance, on which a cooking at constant heat [acts] until they thicken. The bodies are generated from them. Then they transform step by step until gold and silver are made in thousands of years. If these bodies *remained in the mines*, nature would act on them until they would become silver and gold. Anyway, God put into this little underground world, which is like the big world, a wisdom through which it is possible to turn them in one day into silver and gold. They called it sulphur *of the earth. Its preparation* is of three kinds – another interpretation: it is of four kinds. The first one is *calcination* with salt. Then melting, and, after that, congealment and *ceration.* But calcination [...][401] well on a strong fire with salt that takes out from it the sulphur that damages it. Then the bodies will remain pure like calx. He [...][402] Ǧābir ibn Ḥayyān said that they need to calcine those bodies, so that their parts are purified and ready for the *infiltration* of humidity in the vessel [...][403] *which*, if it is like this, one part is able to melt them and increases their dryness, so they are more receptive to humidity and more likely to melt – in my opinion: and they are like gold and silver. One needs to calcine the two [kinds of] tin, lead and iron, because of what we said, and so they purify them from the sulphurs that damage them and they purify from the combustible dross. This is the end of the majority of the bodies. It is done like this to extract from them their humidity which damages

[400] The text of the manuscript presents a *lacuna* here, possibly caused by damage in its *Vorlage.*
[401] The text of the manuscript presents a *lacuna* here, possibly caused by damage in its *Vorlage.*
[402] The text of the manuscript presents a *lacuna* here, possibly caused by damage in its *Vorlage.*
[403] The text of the manuscript presents a *lacuna* here, possibly caused by damage in its *Vorlage.*

fol. 28r

אותם ושבהם יושם הלחות האשי קי מלייורא איניסי לסואה פריפריציאוני לאותו הגוף שצריך ודע זה.

[§ 29] **אבל** אינצירמינטו הגופות זה הוא באמת סוטא לטינטירי החלקים של אותו הדבר אינצירנטי שיעשה דקיק ויותר דק קיפילישק הלחות בגוף אותו שהוצרכנו במלאכת ההשלמה ולא יהיה ריקפיטמינטו אם לא בטחינה ובהשקאה עד שיהיה כשעוה שיותר באש קלה. עשה כל מה שלמדתיך ולא תשנה מצוותי. אך כל האנצירמנטו החכמים החדשים מצאו אותו אבל ההתרה והוא הדיכודיציאוני הוא שתקח אותם הגופות מקולסים אינצירטו ושים אותם בזכוכית וסתום פיו היטב ועשה חפירה אחת ושים בה קש ועמו אותו הזכוכית וכסהו מטיט ושים על הטיט זבל הסוסים ותשליך עליו מים והנח ז״ ימים וזהו התרה בחמימות. אבל התרה בלחות הוא שתעשה גומה א״ ושים בו המזוכך ושים בה מעט מים ותלה עליו הכלי של זכוכית שאמרנו שיש בו האינצירטו שלא יגע המים. וכסה המזוכך במכסה טוב היטב. וסתום הפרק ושים על פי הזכוכית שעוה ועל השעוה חול והנח א״ יום וא״ לילה שלמה ויותר ב״ה

[§ 30] **אבל** ההקפאה זהו. והוא שתקח הזכוכית שיש בה הרפואה המותרת שאמרנו. ושים בקדרה שיש בה אפר סביבה ויהיה מוכבר שלא יטה הכלי והדלק האש תחת הקדרה מעט מעט י״ שעות עד שיוקח ממנו הלחות ויתייבש הלחלוח ויקפא הרפואה כמו צנברו אם יהיה לאדום. ואם הוא ללבן תמצאנו לבן כקנפורא. אחר עשה עמו ותשמח ב״ה

[§ 31] **המאמר** בזהב אלקונדימינטו דאיסו. דע כי הוא אדון הגופות והאבנים היקרים. והוא מלך ולא מן האויר. ולא מן המים. ולא מן העפר. לא יפסד. וגם באש ומן האש ג״כ לא יפסד

fol. 28r

them and which is in them, by adding to the bodies that you wish a fiery humidity *that improves in it its preparation.* Know it!

[§ 29. Ceration of the bodies]
But the *ceration* of the bodies is truly *subject to dyeing* the parts of that *cerated* thing, that makes the humidity in the body thin, and even thinner than hair in the body that is needed for the completed operation. There is no *completeness* if not through grinding and watering [the body] until it becomes like wax that melts on a gentle fire. Do everything that I taught you and do not transgress my prescriptions. This is everything the new sages discovered about *ceration.* But melting, which is the *decoction,* is that you take those calcined and *cerated* bodies, put them in a pot and seal [it] with clay. Make an channel, fill it with straw and with it that cooking pot. Seal its opening well with clay, place horse manure on the clay, pour water over it and let it rest for seven days: this is the melting with heat Melting with humidity is instead that you dig a hole and you place the cooking pot in it. Pour some water over it and hang on it the glass vessel that we have mentioned in which there is the *cerated* [product] so that the water will not be consumed. Close well the cooking pot with a good lid, seal the joint, and put some wax over the opening of the glass vessel and some sand over the wax. Let it rest for one complete day and night and it will melt, with the help of God.

[§ 30. Congealment of the bodies]
But this congealment is that you take a container in which there is the dissolved medicine that we mentioned. Put it in the cooking pot inside and around which there is ash; it is better if the vessel does not tilt. Light a fire under the pot little by little for ten hours until the humidity has been taken out from it, the moisture has dried up, and the medicine has congealed like *cinnabar,* if the operation is for the red. If it is for the white, you will find it white as *camphor.* Then use it and rejoice, with the help of God.

[§ 31. On gold]
Discourse on Gold. *Its preparation.* Know that it is the lord of the bodies and of the precious stones. It is a king that is not corrupted by air, by water or by earth. It is not corrupted even in the fire nor by the fire,

fol. 28v

ולא יחסר אבל מלייורא אנוסו. כי בו באחד לחלוח יתלחלח. וגם הגפרית לא יפסידנו וטבעו הוא דאוגואלי קליטטי. והוא חם. ובו הוא טבעו סודו בענייני הטבעים שלמים בלא טורח ועמל וחסרון בו אם כן עשר חלקים מחמימות וי״ חלקים מקרירות וי״ חלקים מלחות וי״ חלקים מיבשות. ועל כן החכמים אסלטרו אותו שבין כל הגופות האחרים היותר טוב הוא מזוג והיה להם הרכבה ממנו מאקציר הגדול כי הוא אוגייאלי סוסטנציא אדורנטי אמיטינטי לעולם. והוא חלק השמש. ואמרו כי הוא חם ולח דיקונפליסיאוני וכן הזהב בין הגופות כמו השמש בין הכוכבים כי השמש הוא מלך הכוכבים והמאורות. ועמו וישלמו ענייני הארצות. ובעלי חיים. והצומח. והדומם. ועמו יתגדלו וכן הזהב אבסיני בגופות. כי הזהב טיניי כל גוף. ר״ל יצבע. הוא יהיה שאור לאיקציר הלבן והאדום. לא מילייורטו אלא עמו. ולא יושלם עם אחר כמו שהעסה לא תשלם אלא עם השאור שלה וכמו שהגופות בעת שתלבינם ר״ל תנקם. ותעלם. ותוציא מהם הסיד וכל הסרחון תסיר ותרצה לחברם יחדו או לערב אז תשים השאור עם הגופות ותטחון עמם כאחד ותעשה עסה ולישה עם מי דלוסיאו סירום עד שיהיה עסה פירמינטה. ויהיה כעיסה פירמינטה אז אם תאמר איך הוא שאור חגוף. אז עיין ותראה שאור העסה . אם הוא דבר אחר או לא עסה . וזה השאור הוא אם לא גוף . אז א״כ בשם צריך אתה מזא הדבר יהיה לך ותעשה ממנו עם מפתח החכמ״ והפילוסופים. ואמרו כי השאור הוא אותו הדבר שילבין הרפואה. ויסיר ויעמיד העולה שלא יעלה. ויעשה קלים הגופות ויכניס הגופות בגופות ויחברם. ואותו השאור הוא חותם המלאכות. ובעת הכרתו מילייורסי הגופות ב״ה ויושלמו. א״כ אם אותו תניח ותהיה עצל יפסדו המלאכות

fol. 28v

nor diminishes, but it *becomes better in it*, since it by itself becomes humid and damp. Not even sulphur damages it. Its nature is of *uniform quality*. It is hot. Its secret is in its nature, since [its] natures are complete, with no burden, labour or deficiency. So it has ten parts of hotness, ten parts of coldness, ten parts of humidity, and ten parts of dryness. For this reason, the sages *exalted* it because it is the best of all the other bodies, and according to them the composition of the great *elixir*[404] stems from it, because it is of *uniform substance*, *durable and stable* forever. It belongs to the part of the sun. They said that its *composition* is hot and humid, and that gold is among the bodies like the sun is among the stars, since the sun is the king of the stars and of the celestial bodies. With it all the earthly things are completed, the animals, the plants and the minerals, and with it they grow. Gold *draws* the bodies *close* since it *holds* each body – this means: dyes. It is the ferment of the white and red *elixirs* that are not *improved* without it and are not completed with anything else, like dough is not complete without its ferment. Like [with] the bodies, when you whiten them – which means: you purify them –, you sublime them, extract the calx from them, remove all the stench, and you want to unite them in one or mix them, then you put the ferment with the bodies, you grind it with them as one thing, make a dough, and amalgamate with water *of its whey* until it is a *fermented* dough and like a fermented mixture. Then, if you say: "How come it is the ferment of the body?" Consider then and look at the yeast of the dough: is it something else or is it dough? And is this yeast a body or not? So, in the name of God, after this you need to have some of this thing and you need to use it with the key of the sages and of the philosophers. They said that the ferment is the thing that whitens the medicine, that eliminates and stabilises the sublimate so that is does not sublime, that makes the bodies light, that makes the bodies penetrate [other] bodies and unites them. This ferment is the seal of these operations and, when it opens, the bodies are *made better* and are complete, with the help of God. After this, if you rest and are lazy with these things, the operations will be damaged,

[404] In transliteration from the Arabic إكسير in all occurrences in this passage.

fol. 29r

ולא תצלח. כי המשאלהו כמו העסה והשאור וכמו הקיבה לחלב. וכמו המור בין הבשמים. ודע זה כי זהו הזרניך השורף הגופות. ושריפה בזה המקום הוא אשר קונטיני הגופות עם אותו שיתערב באותם שיתערבו עמו מגופות אחרים. אז כשתרצה לשרוף שום גוף מאותם עם אותו אז עם זה ועם התערובה[a] שלו תשרוף אותם.

[§ 32] **קח** זרניך כרכומי חלק וכתוש וטחון עם חומץ שהתרת בו מלח ז״ ימים אחר יבש. וכתוש וטחון ושרוף עם זה כל גוף שתרצה כי ישרף

[§ 33] וידיעת קונדימינתטיו שלו (חוזר לזהב) הוא שתתיכנו בכור הצורפים ושים על ז״ חלקים ממנו א״ חלק מעפרת ואז תכתשנו ותטחננו כי יטחן. ד״א תטחננו עם זרניך קונדיטו כמוהו נ״ל גפרית לאדום. ושים בקדרה קטנה מזוככת וסתום הפרק והדלק תחתיה אש גחלים ותנפח במפוח היטב. ואז עשה ג״ שעות או תשימנו בכבשן הזכוכית. או בתנור של יוצרים לילה א״ ותוציאנו . נ״ל ותרחצנו במים ומלח ואחר במים עד ימתיך ויבש. ואנצירא עם כוכב מוסער עם נשדרא מותר במי אטרמנטו למעט מעט עד שישלם משקלו. ונ״ל להתיר ולהקפיא. אחר שים א״ דרמי על א״ אוק״ בדיל. נ״ל מנוקה ויהיה זהב טוב. ונ״ל להתיר ולהקפיא.

[§ 34] **שער** אחר טוב מזה הרבה. קח מים זכים ושים בם מרתך כתוש יהרתיחהו מעט ואחר חמם טסי זהב דקים ותכבה בם. תתיך וכבה בהם הרבה פעמים עד שיוכל להטחן ואז תשימנו במלאכתך ותשמח עמו

[§ 35] **שער** אחר טוב. קח זהב טחון בלימא וערב עמו זרניך וזה בבגד מהודק עבי וטח סביבו בטיט החכמה. וצלהו א״ לילה עם האש ותוציאנו וכותשנו וטחון היטב. ותרחצנו במי המלח. ושים עמו מלח כתוש היטב כמוהו ושים בכלי ושים בתנור עד שיעשה סיד לבן בע״ה. נ״ל לעשות ההפך אם זהב שרוף עם גפרית ויהיה אדום. ואם נחשת בזרניך ויתלבן.

[§ 36] **שער** טוב למאד. קח זהב טוב כרצונך. ועשה טסין דקים ותחממם באש ותכבם במי המלח. ואחר תחממם ותכבם במים שכבית בהם העפרת ב״כ פעמים עד שהשחיר במים. ותוציאם ר״ל הטסין ותטחנם עד שיהיו אבק רק

---

[a] התערובת

fol. 29r

and you will not succeed, since it is similar to the dough and the ferment, to the stomach for milk and to myrrh among perfumes. Know it, since this is the *arsenic*[405] that burns the bodies. Combustion [happens] in the place that *contains* the bodies with what mixes with them, and they mix with it from other bodies. So, if you wish to burn with this any body among these, then burn them with it and with a compound of it.

[§ 32. Preparation of arsenic]
Take a part of yellow *arsenic,* pulverise it, and grind it for seven days with vinegar in which salt has been dissolved. Then dry [it]. Crush, grind and burn with this any body you wish because it will burn.

[§ 33. Preparation of gold (i)]
The knowledge of its *preparation* (it goes back to gold) is that you melt it in a goldsmith's furnace. Put one part of lead over seven parts of it. Then crush it and grind it since it will be ground – another thing: grind it with as much *prepared arsenic,*[406] – in my opinion: sulphur for the red. Put it in a small glazed pot, seal the junction, light under it a fire of coals and blow on it well with bellows. Do this for three hours or put it in a glass furnace or potters' oven for a night. Then take it out – in my opinion: wash it with water and salt and then with water until it melts and dries. *Cerate* it with mercury sublimed with dissolved sal ammoniac[407] in water of *atrament* little by little until its weight is complete – In my opinion: melting and congealing. Then put a *dirham*[408] [of it] over one *ūqiyya*[409] of tin – in my opinion: purified – and it will be good gold – in my opinion: melting and congealing.

[§ 34. Preparation of gold (ii)]
Another **chapter** much better than this. Take clean water and put ground *litharge*[410] into it. Boil it slowly, then warm up some fine sheets of gold and quench them in it. Melt and quench many times until it is possible to grind them. Then put it among your operations and rejoice with it.

[§ 35. Preparation of gold (iii)]
Another good **chapter.** Take gold ground into *filings* and mix with it *arsenic.* [Do] this in a thick and tight cloth, and cover around it with the clay of wisdom. Roast it for one night with fire, take it out, crush it, and grind it well. Wash it with water of salt. Put with it the same amount of pulverised salt and put it in a vessel. Place it in an oven until it becomes a white calx, with the help of God – in my opinion: doing the opposite, if gold is burnt with sulphur: it will be red. And if lead [is] in *arsenic,* then it becomes white.

[§ 36. Preparation of gold (iv)]
Another very good **chapter.** Take as much as you want of good gold and make [with it] thin sheets. Warm it up on a fire and quench it in water of salt. Then warm them and quench them in water in which lead has been quenched twenty-two times until it has darkened in the water. Take them out – it means: the sheets – and grind them well until they are a fine dust

[405] In transliteration from the Arabic زرنيخ in all occurrences in this section.
[406] This phrase combines a transliteration from the Arabic زرنيخ ("arsenic") and a transliteration from the Italian "condito" ("prepared").
[407] In transliteration from the Arabic نشادر in the text.
[408] In transliteration from the Arabic درهم in the text.
[409] In transliteration from Arabic أوقية in the text.
[410] In transliteration from the Arabic مرتك in the text.

fol. 29v

היטב עד שיהיה כשחק מאזנים ואחר עשה מלאכתך עמו ב״ה (נ״ל והוא שתקח כוכב מוסער ג״ כמוהו וגפרית מלובן מאודם באשו ואנצירא עם מי נשדרא שהתרת בו הזאג ותתיר ותקפיא ז״פ).

[§ 37] **ד״א** קח מרתך וכתוש ולוש עם שמן ושים בחרס הצורפים עד שתחתית הכלי יהיה מכוסה. ואחר קח זהב טחון בלימא ושים באלביכו *נ״ל באלביט* ותתיכנו וכשיתקרר טחון אותו ותשמור לעת הצורך.

[§ 38] **ד״א** קח מן הסידים הללו ממה שתרצה א״ חלק אחר שהוא מקולס וטחון אותו עם כמוהו נשדרא מותר עד שישתהו כלו. נ״ל זג מאודם מותר וסנן ותתיר בו נשדרא מותר ושים בכלי זכוכית עם חומץ חד ביותר. ושים במקום לח עד ז״ ימים ויותר או עד שיותר כי יותר למים אדומים כדם. אז אנצירא ~~עמו~~ עם זה המים כוכב מוסער שסערת עם מלח והרבה פעמים תטחון ותשקה ויבש עד שישלם היטב ואחר תתירנו. וכשיותר תקפיאנו ושים א״ דרמי על נ״ מכסף נ״ל ק״ ויהיה זהב ב״ה

[§ 39] **המאמר** רציאו על הכסף. דע כי הכסף הוא גוף טהור אך פחות מן הזהב והוא מחלק לבנה כי הלבנה הוא מאיר קטון מן השמש ומשקלו הוא משקל פחות ממשקל הזהב. וכן ערך הכסף על ערך הזהב ואמנם הכסף יפסד בלחות ובמקום לח וטעמו הוא חמוץ. יקרו מקרים רבים ובגפריתים. ויפסדו במקרים. ובאש יזוכך וזולת הזהב אין גוף טוב ממנו והוא יותר ההוא לזהב מכל הגופות ונסתרו זהב כגלו כסף והזהב נסתרו כסף וכגלהו זהב וטבעו קר ולח ואמרו אחרים קר ויבש ויתערב בזהב ובנחשת ומקבל הצבע גם עשינו ממנו אקציר לבן ואדום אחר עשינו ממנו זהב והרבה מקרים רעים יוקרו ובמהרה לפי שהוא זך הרבה[411] ואינו סבלן לאותן המקרים שיקרו לו והגפריתים והבדילין. והבדילין הם שונאים לו ויעשו לו רעה ואינם ראוים לו אבל

חלום אמר לי פלורינטינו קח זהב טחון או בלימא או מלומא לאחר שהוצאת הכוכב ממנו ותשקה וצלה מי הזג ומי נשדרא עד שישתה כמוהו

[411] Blank space in the manuscript, possibly due to a corruption of the *Vorlage*.

fol. 29v

until they are like powder of the ears.[412] After that, perform your operations with it, with the help of God. (In my opinion: it is that you take three times its quantity of sublimed mercury and whitened [or] reddened sulphur in its fire and *cerate* [it] with water of *sal ammoniac*[413] in which *vitriol*[414] has been dissolved. Melt it and congeal it – this is a commentary).

[§ 37. Preparation of gold (v)]
Another. Take *litharge*,[415] pulverise it and knead it with oil. Put it in a goldsmith's crucible until the bottom of the vessel is covered. Then take gold ground into *filings*, put it in an *alembic* *in my opinion: in an egg*, melt it and, when it has cooled down, grind it and save it for the time of need.

[§ 38. Preparation of gold (vi)]
Another. Take one part of the one you chose among those calces after they were calcined, and grind it with the same amount of dissolved *sal ammoniac* until it has drunk it all – in my opinion: reddened *vitriol*. Filter it and dissolve in it dissolved *sal ammoniac*. Put it in a glass vessel with more sharp vinegar. Put it in a humid place for seven or more days or until it is dissolved, since it will dissolve into a water red like blood. Then with this water *cerate* some sublimed mercury that you have sublimed with salt. Pulverise it many times, water it and dry it until it will be completed well; then dissolve it. When it is dissolved, congeal it and put one *dirham* on fifty *dirham* of silver – in my opinion: one hundred – and it will be gold, with the help of God.

[§ 39. On silver]
The discourse of *the method* for silver. Know that silver is a pure body, but less [pure] than gold. It is from the part of the moon since the moon is less luminous than the sun. Its weight is lower than that of gold, and the composition of silver is similar to the composition of gold. Indeed, silver is corrupted in humidity and in a humid place and its taste is sour. Many accidents happen [to it] and with sulphurs and it is corrupted by the accidents. It is purified with fire. Apart from gold, there is no better body than it and it is the closest to gold among all the bodies. What is hidden in gold is apparent in silver and what is hidden in silver is apparent in gold. Its nature is cold and humid; others said that it is cold and dry. It mixes with gold and with copper and receives the dye. Also from it we made the white and red *elixir*.[416] Then from it we made gold. Many bad accidents happen to it and very quickly, since it is very pure [...][417] It does not endure those accidents that happen to it nor [does it endure] sulphurs and [the kinds of] tin. [The kinds of] tin are enemies to it and they damage it: they are unsuitable for it, but
*Dream. The *Florentine* told me: take ground gold or *filings* or an *amalgam* with another [ingredient] from which mercury comes out and water it, and roast water of *vitriol* and water of *sal ammoniac* until it has drunk as much as its amount*

[412] This peculiar expression has no parallels in the treatise and may be the result of a corruption of the text that took place in the process of its transmission or translation.
[413] In transliteration from the Arabic نشادر in all the occurrences in this section.
[414] In transliteration from the Arabic زاج in all the occurrences in this section.
[415] In transliteration from the Arabic مرتك .
[416] In transliteration from the Arabic إكسير .
[417] The text of the manuscript presents a *lacuna* here, possibly caused by damage in its *Vorlage*.

fol. 30r

הגפריתים לבד הן קונדימינטו של כסף ושל עופרת ושל נחשת. והזרניך הוא קונדימינטו לזהב ולקלעי ולבדיל וגם לעופרת ודע זה. והכסף הוא גוף שלם. ר״ל מאדימים הלבן הגדול. ר״ל עשיית המלאכה הגדולה. ויעשה ג״כ הקלעי אך עשיה פחותה. והזרניך והכוכב הם נפש ורוח של אקציר. והנשדרא הוא השמש שלבן *נ״ל שלהן* והוא המחבר אותן. ר״ל הדברים של אקציר.

[§ 40] לפמא דלקונדימינטו שלו הוא שהכסף תתיך באלביט ושים. עמו כמוהו קלעי. כשיתקרר תטחננו ושמור.

[§ 41] **ד״א** תתיך קלעי אחר וכבה אותו פעמים במי המלח אחר תחמם ר״ל טסין. של כסף וכבה במים ההם אחר ששמת בכל ליט״ של מים ההם א״ דרמי מגפרית ומזרניך. ואז אותו הכסף יתקלס וישחיר אז עשה ממנו עם המלח המותר וצלה בתנור מודלק אחר תרחצנו ויהיה לבן וטהור.

[§ 42] **ד״א** אם תבשל הגפרית במים ומלח ותתיך כסף טחון ותכבה במים ההם ואחר שים בכבשן לילה א״ אחר תרחץ במים היטב ותנקהו מן המלח עד ימתיק ואז יהיה קלס ב״ה.

[§ 43] **ד״א** קח כסף טחון א״ חלק וכתוש וטחון ושים בבגד ושים בכור ושים על הגחלים כשיתחיל להתחמם השלך עליו מעט גפרית עד שישרף ויעשה אפר אז תטחננו עם המלח מותר כמו שאמרנו ויהיה עשוי אפר לבן ב״ה. נ״ל כי יהיה כרכומי ולא לבן.

[§ 44] **ד״א** קח טסי כסף דקים למאד ושים על הגחלים עד יאדימו באש ותכבם בשמן כן עשה עד ישחיקו הטסין. אח״כ תטחון במלח המותר הנזכר למעלה ושמור. אחר קח מזה חלק נ״ל וחלק כוכב וכמוהו זרניך נ״ל אם תוסיף חלק כוכב מוסער מעל הכסף ז״פ ואנצירא ותתיר ותקפיא ותשליך על מאה בדיל וא״ בדיל על נ״ נחשת ב״ה . ואנצירא עם אלומי מותר בשתן

fol. 30r

only the sulphurs are the [ingredient for the] *preparation* of silver, lead, and copper. *Arsenic*[418] is the [ingredient for the] *preparation* of gold, of the *qalaʿī*,[419] of tin, and of lead as well. Know this! Silver is a perfect body – this means: they redden the great white – this means: the making of the great operation. The *qalaʿī* acts in this way, but it is weaker. *Arsenic* and mercury are the soul and the spirit of the *elixir.*[420] *Sal ammoniac*[421] is the sun that whitens *in my opinion: their [sun]* and their unifier – this means: the things pertaining to the *elixir.*

[§ 40. Preparation of silver (i)]
The account of its preparation is that you melt silver in a *crucible*[422] and put with it the same amount of *qalaʿī.* When it cools down, grind it and save it.

[§ 41. Preparation of silver (ii)]
Another. Melt some other *qalaʿī* and quench it few times in water of salt. After that, warm them – this means: the sheets of silver – and quench them in the same water, after adding one *dirham* of sulphur and *arsenic* for every *pound* of that water. So this silver is calcined and has become black. Do this with it with dissolved salt and roast it in a fiery oven. Then take it out: it is white and pure.

[§ 42. Preparation of silver (iii)]
Another. If you cook sulphur in water and salt, you melt ground silver, you quench it in the same water, then you put it in an oven for a night and then you wash it well in water and you clean it from the salt until it is sweet, then, with the help of God, it will be calx.

[§ 43. Preparation of silver (iv)]
Another. Take one part of ground silver, pulverise it, grind it and put it in a cloth. Put [the cloth] in an oven and place it over hot coals. When it begins to warm up, pour over it a little quantity of sulphur, until it burns and becomes ash. Then grind it with dissolved salt, as we said: it will become a white ash, with the help of God – in my opinion: since it will be yellow and not white.

[§ 44. Preparation of silver (v)]
Another. Take some very thin silver sheets and put them on coals until they become red in the fire. Quench them in oil. Do this until the sheets are pulverised. After that, grind them in the dissolved salt mentioned above and put it away. Then take one part of it – in my opinion: one part of mercury and an equal part of *arsenic* – in my opinion: if you add one part of sublimed mercury over the silver – this is a commentary – *cerate* it, melt it, congeal it and pour it over one hundred [parts of] tin; pour one part of tin over fifty [parts] of copper, with the help of God. *Cerate* with *alum* dissolved in urine

[418] In transliteration from the Arabic زرنيخ in all occurrences in this section.
[419] In transliteration from the Arabic قلعي in all occurrences in this section.
[420] In transliteration from the Arabic إكسير in all occurrences in this section.
[421] In transliteration from the Arabic نشادر in all occurrences in this section.
[422] In transliteration from the Arabic بوط .

fol. 30v

נערים עם כמוהו נשדרא. ואנצירא עם טחינה והשקאה ויבשות עד שיהיה אנצירטו היטב ויותך כשעוה. אז שים כל זה בקליפת ביצה וטח אותה סביבה היטב ועליה שים קליפה אחרת וטח הפרק היטב בטיט החכמה ויבש. ואחר שים אותו באש זבל השוורים או בבתייא שלא תהיה חמה ביותר אלו מזוגה א״ לילה וא״ יום. אחר תוציאנו ותמצאנו כסף קפוי השלך ממנו א״ דרמי על ח״ דרמי מבדיל מנוקה. נ״ל או על נחשת או על הברזל מנוקין ויהיה כסף טוב ויפה ב״ה.

[§ 45] **שער** אחר טוב מזה מאד. קח א״ אוקיי״ טסי כסף ומזרניך קונדיטו ששורף ומקלס הגופות רביע אוקיי״. ושים שורה שורה מן הטסין ומן עפר הזרניך בקליפת הביצה וטח בטיט סביבה. ושים בקדרה שיהיה בה אפר מוכברת והזהר שלא תטה הביצה ושים הקדרה בתנור הלחם א״ לילה וא״ יום ותוציאנו. ואנצירא אותו במי אלומי ונשדרא עד שיהיה כשעוה אז שים מזה על הבדיל ויהיה כסף טוב ב״ה. אבל אם תשים הכסף הטחון עם הזרניך ותטחנם. ותשים אותם לישרף יהיה מלאכתך יותר טוב. ותשליך על כל מה שתרצה. נ״ל לכוכב. ותאוותך ישלם כי דומה לקיבה בחלב. נ״ל להבין מסוף שערי הברזל לא בוכר. כי זה יקפיא הכוכב באנפולה אטורטא וזה ילבין עוד כונה.

[§ 46] **המאמר** בברזל. הברזל הוא מחלק מאדים. וטבעו חם ויבש ואמרו כי הוא קר ויבש. והוא זכר ונקבה. חמוץ בטעמו והוא קשר סובל האש הרבה. ולוחם עם האש . וד״ דברים מתיכין אותו. הזרניך העפרת והמגניסיאה ומרקשיתא. ובעת שעם הזהב יתחבר לא יפרד ממנו לעולם. לא בעבור רובץ ולא בעבור דבר אחר. ואם תערב אותו עם כסף וזהב בעת שהכסף והזהב יהיו מחברים יחד יצבעם צביעה טובה ונאה. ויהיה יותר טוב אם יתבשל. וזהו סוד גדול בסודי היאודים. והחכמים בין כל הגופות

fol. 30v

of young boys in the same amount of *sal ammoniac*;[423] *cerate* it by grinding, watering, and drying it until it is well *cerated* and it melts like wax. Then put all this in an eggshell, cover it well all around and put over it another shell. Cover the joint well with clay of wisdom and let it dry. Then put it for one night or one day in a fire of ox dung or in a *bottle* that does not become hotter than this mixture. Then take it out and you will find it [like] solid silver. Put one *dirham* on eight *dirham* of pure tin – in my opinion: or on pure copper or iron – and it will be good and beautiful silver, with the help of God.

[§ 45. Preparation of silver (vi)]
Another **chapter** much better than this. Take one *ūqiyya*[424] of sheets of silver and a quarter of an *ūqiyya* of *prepared arsenic*[425] that burns *and calcifies the bodies*. Then put one layer after the other of the sheets and of dirt in an egg. Cover around it with clay. Bury it in a pan in which there is ash; make sure that the egg is not tilted and put the pan in a bread oven for one night and one day. Take it out. *Cerate* it with water of *alum* and *sal ammoniac* until it is like wax. Then put [some] of it on tin: it will be good silver, with the help of God. But if you put ground silver with *arsenic*, you grind them and you make them burn, your operation will be much better. Pour it on everything you want – in my opinion: on mercury – and you will be completely satisfied, because it is similar to the stomach for milk – in my opinion: it is understood from the end of the chapters on iron that it is not preferable, because this congeals mercury in a *twisted vial*, and it will whiten again.

[§ 46. On iron]
Discourse on iron. Iron is from the part of Mars, and its nature is hot and dry. They said that is cold and dry. It is male and female and its taste is sour. It is hard and very resistant to fire and it fights with fire. Four things melt it: *arsenic*, lead, *magnesia*, and *marcasite*. When it binds with gold, it will never separate from it, neither stretching it out nor in any other way. If you bind it with silver and gold, when silver and gold are united together, it dyes them with a good and beautiful dye. And this is even better if it is cooked. This is one of the great secrets of the Jews.[426] The sages, among all bodies,

[423] In transliteration from the Arabic نشادر in all occurrences in this section.
[424] In transliteration from the Arabic أوقية in all occurrences in this section.
[425] In transliteration from the Arabic زرنيخ in all occurrences in this section.
[426] Only the Hebrew text identifies these sages as Jews; all other versions present them as Indian sages.

fol. 31r

בחרו הברזל מפני שבמהרה ירכיב עשייתו ובקלות. וצביעתו הוא במלאכתנו קיימת בלבן ובאדום. והקונדימינטו שלו הוא להתיכו בהתרה. ובנשדרא. ואנחנו נזכיר אח״כ אלו השמות.

[§ 47] ידיעת קונדימינטו שלו הוא מה שאמרנו כבר אמרנו כי הברזל הוא יותר קשה וחזק מכל הגופות. ולא יותך באש אם לא בכח גוף חזק. וזהו שתקח טסי ברזל דקים וחמם באש עד יאדימו ותכבם בשמן באותו שהעפרת יהיה מותך בו הרבה פעמים עד יחסר וינקב. והוא שיעשה ממנו חתיכות שיהיו ג״כ קלים אז תשברם ותכתשם. ס״א תתיך אחר כן העופרת ההוא או אחר. ושים עליהם כוכב כמוהו וטחון יחד עד שיעשה אבק ועשה שטח מאותו האבק. ואחר תשים עליהם טסי ברזל שורה ממנו ושורה מן האבק עד שיתמלא כל הכלי פה לפה. אז תשים בתנור והדלק האש תחתיו ותנפח במפוח עד שיותך. אחר תוציאנו כן עשה עד שייטב בעיניך דמותו ורכותו ואז כשיהיה כך אז תעשה עמו ב״ה. נ״ל כי אז ילבין כל הגוף וירפאם ויקפיא הכוכב

[§ 48] **ד״א** קח טסי ברזל הנדי טחון כרצונך. וזרניך כתוש היטב רביעיתו. וערב

נ״ל כי כל א׳׳ מאקצירי ברזל יקפיא ויצבע הכוכב א׳׳ על מאה בקירוב

היטב ושים בכלי עד ימלא. וטח הפרק והכלי מטיט החכמה ויבש. ושים בתנור והדלק האש ג׳׳ שעות ותנפח במפוחים באש חזק עד שיותך. אז תוציאנו ותקרר. וטחון עם מלח אלקלי מעורב עם שמן. וזה יהיה ששיתו ר״ל הסל אלקלי ושים בבוט. והדלק תחתיו אש הגחלים. ונפח במפוח עד שירד למטה המטיריאה והסיג ישאר למעלה ממנו. אחר תוציאנו והחזר עוד הדבר פעמים עד שיותך כעופרת. ואם תרצה עוד תקח מנטרון וטנכר ס״א נטרו או מלח זכוכית. וטחון אותו יחד ולוש עם שמן. ומאותו עשה פלולים ותטעים אותו הברזל מהם . ותתיך אותו י״פ אז יותך מהרה ויהיה יותר לבן מבתחלה וכל כך פעעמי״ תתיכנו עד שיותך כעפרת ויהיה רך תחת הפטיש ולא ישבר אז תקח מזה הברזל מתוקן א׳׳ דרמי והשלך על ה׳׳ דרמי מקלעי מנוקה ס״א על ט״ו. עד שיותך ויצא כסף טוב להשליך ולקבל הפטיש ב״ה נ״ל אם

fol. 31r

chose iron, because its preparation is composed quickly and easily. Its colour is stable through our operations for the white and for the red. Its *preparation* is that you melt it by dissolving and with *sal ammoniac*.[427] In what follows we are going to mention these names.

[§ 47. Preparation of iron (i)]
The knowledge of *its preparation* is what we have said. We have already said that iron is the hardest and the strongest among all bodies, and that it does not melt in the fire, if not with the quality of a strong body. [Its preparation] is that you take thin sheets of iron. Warm them in the fire until they become red. Quench them multiple times in oil in which lead has been melted, until they shrink and become pierced. This requires also that you make in this way pieces [of iron] that are also light. Then break them and pulverise them – another opinion: after this melt that lead or another. Put over them the same quantity of mercury, grind them together until they become dust. Make a layer of that dust and then put over it a layer of sheets of iron and a layer of dust until the vessel is completely filled up. Then put it in an oven, light the fire under it and blow with the bellows until it melts. After that, take it out. Do like this until its colour and softness appear fine to you. When it is like this, then use it, with the help of God – in my opinion: because it whitens and heals every body and congeals mercury.

[§ 48. Preparation of iron (ii)]
Another. Take as many sheets of ground Indian iron as you want and a quarter of that quantity of well-pulverised *arsenic*.[428] *In my opinion: because every part of the elixir[429] of iron coagulates and dyes mercury, one part for about one hundred parts* Mix them well and put them in a vessel until it is filled. Cover the joint and the vessel with the clay of wisdom and let it dry. Put [it] in the oven, light the fire for three hours and blow on the strong fire with bellows until [the contents of the vase] melt. Take it out and let it cool down. Grind it with *alkali*[430] salt mixed with oil, and this should be one sixth [of the other] – this means: *alkali* salt and put it in a crucible.[431] Light under it a charcoal fire and blow with the bellows until *the matter* descends to the bottom and the dross remains over it. Then take it out and repeat this work multiple times until it melts like lead. If you want, take again *natron*[432] and *borax*[433] – another interpretation: *natron* or salt of glass. Grind them together, knead [them] with oil and make with it *small balls*. Feed the iron with them. Melt it ten times: then it will melt quickly and will be whiter than it used to be at the beginning. Melt it in this way multiple times until it melts like lead, becomes malleable to [the blows] of a hammer and does not break. Then take one *dirham* of this rectified iron and pour it on five *dirham* of cleaned *qalaʿī*[434] – another interpretation: on fifteen – until it melts and it comes out as good silver that can be poured and worked with the hammer, with the help of God – in my opinion: if

[427] In transliteration from Arabic نشادر .
[428] In transliteration from the Arabic زرنيخ .
[429] In transliteration from Arabic إكسير .
[430] In transliteration from Arabic القلي .
[431] In transliteration from Arabic بوط .
[432] In transliteration from Arabic نطروب .
[433] In transliteration from Arabic تنكار in the text.
[434] In transliteration from the Arabic قلعي in the text.

fol. 31v

תשליך חלק מזה הכסף על הנחשת יהיה הכל כסף. אם תרצה לקח[a] עפרת טחון. ותתיך ברזל טחון עמו כמו שעשית מן הזרניך. אז אותו יותך מהרה ויתוסף לו לבנות ורכות ב״ה.

[§ 49] **ד״א** קח אלקלי כרצונך. ולוש עמו זרניך כרכומי. נ״ל גם קליפות ביצה שרופים וטח עמו טסי הברזל וחממם היטב ותכבם כ״כ פעמים עד שיתלבנו כדמות כסף וידמה לכסף בכל עניין. אז תוציאנו כי הוא כסף זך חזק והוא טוב. ואין הפרש בינו ובין הכסף. עוד תטחנו ושים באלביט ושים עליו מדיתריירו מלובן. או מן הצירוסי או מן חתיכת מנטרו של חכמים פי״ הוא ממלח בורק. או שים דבר שדומה לנטרו. או הוא נשדרא אז יותך מהרה כמים. וכשיותך אז תנקהו כן תעשה הרבה פעמים עד שייטב בעיניך הדמות שלו ויותך מהרה כעופרת. אחר תוציאנו וכתוש אותו עם אלקלי נקי. וערב עם כסף טוב ועשה עמו מה שתרצה. נ״ל עם קלעי או כוכב כי אין טוב ממנו בכל העולם. כי הוא עשיה במהרה למאד ובזמן קצור ישלם אך טורח יש בו. אז כשהברזל יתקלס הוא טורח גדול. ואין יותר טוב אקונדימנטו

[§ 50] **ד״א** קח שילפו כרכומי חלק. וכתוש אותו ושים בחרס וכשירתח בשמן כפלים כמוהו עד שיותר בו השולפו. אחר קח טסי הברזל כמוהו ותחמם באש עד יאדימו ותיכבם בשולפו ובשמן הנזכר כן תעשה הרבה פעמים עד שיתקלסו הטסין. ואז יהיה שחור וטחון. אם תרצה קח הטסין ומשח אותן מן הגפרית לא מותר ושים בקדרה מזוככת וכסה אותם במכסהו וטח הפרק מטיט החכמה ושים בתנור הלחם א״ לילה. אך תקברנה בגחלי אש גדול וטחון היטב. נ״ל ורחץ ג״כ . אחר קח אטרמנטו כמו הברזל וצלה בתנור הלחם בקדירה חדשה עד יאדים אחר תוציאנו ותתירנו בשמן *נ״ל בשתן* ישן נערים ר״ל שעמד ימים. אחר שהותר סנן במזוכך ויתיר

חלום טיסטוייני יש בו מלה וחברתי לקליפה ולשמרים. אם כן תוציא חמימות[b] וכו״ל

---

[a] ליקח [b] המימות

fol. 31v

you pour one part of this silver on copper, it all becomes silver. If you want to take ground lead and melt ground iron with it as you did with *arsenic*,[435] then it melts it quickly and its whiteness and softness will increase, with the help of God.

[§ 49. Preparation of iron (iii)]
Another. Take as much as you want of *alkali*,[436] knead yellow *arsenic* with it – in my opinion: also burnt eggshells – and cover with it some sheets of iron. Warm them well and quench them as many times as needed until they whiten to resemble silver: they will be similar to silver from all points of view. Then take them out, because it is pure, strong and good silver and there is no difference between it and silver. Grind it again and put it in a *crucible*;[437] put on it some whitened *litharge*[438] or some *ceruse* or a piece of *natron* of the sages, which derives from *borax* salt, or put something similar to *natron* or *sal ammoniac*.[439] Then it will melt like water. When it is melted, clean it. Do this many times until its colour looks fine to you and it melts quickly like lead. After that, take it out and pulverise it with pure *alkali*. Mix it with good silver and do with it what you want – in my opinion: with *qalaʿī*[440] or mercury – because there is nothing better than it in the whole world, since this is done with a very quick action and it is completed in a short time. There is anyway an effort in it, because it takes a great effort for iron to be calcined, but no *preparation* is better than this.

[§ 50. Preparation of yellow sulphur]
Another. Take one part of yellow *sulphur*, grind it and put it in a piece of pottery. When it boils in double its quantity of oil until the *sulphur* will melt in it,[441] then take the same amount of iron sheets and warm them in the fire until they become red. Quench them in the aforementioned *sulphur* and oil. Do this many times, until the sheets are calcined: they will be dark. Grind. If you want, take the sheets, smear them with undissolved sulphur, and put them in a glass pan. Close it with its lid, seal the joint with the clay of wisdom and place it in a bread oven for one night. So bury it in the ashes of a big fire – in my opinion: wash it as well. After that, take as much *atrament* as the quantity of iron, roast it in a bread oven in a brand new pan until it has become red. Then take it out and dissolve it in aged oil – in my opinion: in urine – of young boys – this means: that has been left for days. After it has melted, filter it in a pan and it will melt [...]

*Dream of *tysṭwyyny*.[442] In it there is salt and I mixed it with the eggshell and the ferments. If so, take out the warmth*[443]

[435] In transliteration from the Arabic زرنيخ in all occurrences in this section.
[436] In transliteration from the Arabic القلي in the text.
[437] In transliteration from the Arabic بوط in the text.
[438] In transliteration from Italian in the text. The copyist provides a transliteration of difficult interpretation, probably caused by a problematic reading in the manuscript from which he was copying this section.
[439] In transliteration from the Arabic نشادر .
[440] In transliteration from the Arabic القلي in the text.
[441] The syntax of this passage appears corrupted.
[442] On the interpretation of this word in transliteration, see *Introduction*, § 5.7.
[443] The marginal note is concluded by an abbreviation that I am unable to solve.

fol. 32r

[§ 51] ואם תרצה קח מן הרפואה חלק וב״ חלקים מכסף. ועשה ממנו טסין דקים אחר שים בטסין מן הרפואה. ר״ל משח אותם מב״ עבריהם ושים באלביט ויבשם כלם פירי אנטריבלי היטב. ואחר תניח לקרר. אחר תשקול אותו כלו ולא תמצא יותר מן המשקל הראשון של כסף ודמותו יהיה ציטרינו ונאה אחר תתיכנו ב״ או ג״ פעמים ותתיך הכסף בשתן נערים ישן. אז פניו יהיה כרכומי. אז שים ממנו א״ חלק עם חצי חלק זהב ויהיה טוב.

[§ 52] **קונדימינטו** אחר. קח נחשת שרוף ב״ חלקים וטחון אותו היטב עם א״ חלק שולפו במכתש נחושת ושים בקדרה מזוככת וכסה וטח היטב הפרק. ושים בתנור הלחם ג״ שעות או ד״ ואם במקום נחשת שרוף זנייר טוב וזך תרצה לשים אותו שתשרוף עם הגפרית כמו שאמרתי לך. נ״ל ורחץ היטב עד ימתיך אחר תטחון עם מי אטרמנטו ונשדרא כמו שתדע. אז שים א״ דרמי על ב״ דרמי מכסף. ויהיה זהב טוב ויפה. ואם לא ייטב בעיניך דמותו תתיכהו פעם אחרת ושים עליו א״ דרמי אחר מכסף *נ״ל מן הרפואה*. ואז יהיה יותר טוב ותודה לשם. וזה ג״פ תוכל לעשות ביום א״ ואם תרצה לעשות שישוה לאדום קח מן האחד ומן האחר בשוה ותתיכם ג״פ ואז יהיה אדום טוב ויפה.

[§ 53] **נקוי** הברזל והנחשת. נ״ל גם בדיל ועופרת. קח שומים הרבה ד״ ראשין. או כרצונך וכתושם במכתש של עץ עם מעט מלח ועם אלומי ואז שים בכלי ואחר תשים חומץ חד עד שתתכסה כל השומין ותבשל היטב. וכבר שמנו בדסיזיאוני אחר מי סאבון והוא טוב למאד ויפה מכל אותם הדברים שעמו סִאוֹפְירנו. אבל אני לא נסיתי. אחר קח טסי נחשת. נ״ל או ברזל דקים למאד ותחממם ותכבה בם הרבה פעמים עד שיתלבנו. אז ערב באותם הלבנים. אם ייטב בעיניך ואם

fol. 32r[444]

[§ 51. Preparation of copper (i)]
[...] If you want, take one part of medicine and two parts of silver. Make thin sheets with it and, after that, put some medicine on the sheets – this means: smear them on both sides. Put them in a *crucible*[445] and dry all of them well *in an antribulo*.[446] Then let them rest to dry and weigh it all: you will not find more than the original weight of silver and its colour will be *citrine* and beautiful. Then melt it two or three times and melt the silver in aged urine of young boys: its surface will be yellow. Then put one part of it with half part of gold: it will be good.

[§ 52. Preparation of copper (ii)]
Another ***preparation.*** Take two parts of burnt copper, grind them finely with one part of *sulphur* in a mortar [made] of copper, and put them in a glass pan. Close it and seal the joint well. Put it in a bread oven for three or four hours. If, instead of burnt copper, you want to use good and pure *verdigris*, [you need to] burn it with sulphur, as I have already told you – in my opinion: and wash it well until it is sweet. Then grind it with water of *atrament* and *sal ammoniac*,[447] as you know. Then put one *dirham* of it on two *dirham* of silver: it will be good and beautiful gold. And if its colour does not look good to you, melt it another time and put on it another *dirham* of silver – in my opinion: of the medicine. Then it will be better. Thank God. This [preparation] can be performed three times in one day. If you want to make it similar to red, take similar quantities of the one and of the other and melt them three times: it will be red, good and beautiful.

[§ 53. Cleaning of iron and copper]
Cleaning of iron and copper – in my opinion: also of tin and lead. Take four heads of garlic or as much as you want and pulverise them in a wooden mortar with some salt and with *alum*. Then put it in a vessel and then pour over it sharp vinegar until it covers all the garlic. Cook it well. We have already mentioned it in another *description* of impure water, and it is much better and more beautiful than all the other things with which *we operate*, but I have not tried it. After that, take very fine sheets of copper – in my opinion: or of iron – and warm them up and quench them in these many times until they whiten. Then mix with it the white ones. If it looks good to you, [fine]; and if

[444] The text of the Hebrew MS is interrupted abruptly at the end of folio 31v. The opening of the next folio, which is numbered progressively in Arabic numerals as 32, does not contain the continuation of the previous recipe but rather the end of one of the recipes devoted to the preparation of copper. In the Hebrew MS, the theoretical discourse on copper and its features and the first recipes devoted to this metal have not been preserved. It is very likely that this extensive lacuna corresponds to a folio of the manuscript that was lost before the trimming and the numbering of the pages in Arabic numerals.

[445] In transliteration from Arabic بوط .

[446] In transliteration from Italian in the text. The original word is difficult to reconstruct, but, given the context and the other versions, I believe a mortar is here intended. The word "antribulo" could derive from the Italian verb "tribbiare/trebbiare" with the meaning of "grinding."

[447] In transliteration from Arabic نشادر .

fol. 32v

תרצה חתכם במספרים ושים באלביט עם מעט טנכר או נטרו הוא אבן מלוח. אז כשיהיה מותר *נ״ל מותך* שים א׳׳ דרמי מאותו לבד. נ״ל הברזל או הנחשת עם א׳׳ דרמי מאקציר לבן. ותוסיף א׳׳ דרמי מכסף ויבא לך כסף לבן שיותך ככסף ויסבול האש ככסף ויעמוד בפטיש ולא ישתנה ב״ה.

[§ 54] **ד״א** יפה וטוב. קח לימטורא של נחשת א׳׳ ליטרא. וכמוהו כוכב נ״ל בדיל. ס״א ד׳׳ כמוהו. וטחון היטב פירי אנטריבלו עד שיהיה כעסה אחר תרחץ היטב עם מלח וחומץ עד שיתנקה היטב. ואחר עם מים זך תרחץ אחר תיבש. וקח אטרמנטו א׳׳ ליטרא ותתיר בכפלים שתן נערים ישן וסנן. ושים עליו א׳׳ ליטרא נשדרא. נ״ל השקות וצלות. ואינצירא עם זה המים למעט מעט עד שישתהו כלו. ויהיה אותו צבוע מדמות בין כרכומי לאדום הנקרא בלשון ערב אצפר. (נ״ל משח עמו טסי כסף מנוקה או כסף ובדיל וחמם מעט מעט עד יכנס בו ועשה עד יהיה זהב. אז תתיך ושים עליהם) כבר חברנו הכוכב עם הנחשת אז עשה עמו ותשמח. נ״ל או לצלותו להקפותו בכדור זכוכית מלא מטויין ולהשקותו עד יקפה ולא יעשן. או להתיר ולהקפיא עד תגיע. או להדליף. או להקפות בכור מכוסה מכור אחר ומטויין בדרך יבר בצביעת הלבן לאדום. או למלא אנפולא ותקיפא באש הסובין או באפר חם. או תקפיא ותתיך. ואם תרצה ללבן . במקום זאג שים אלומי יימיני ועשה עמו כמו שאמרנו שתשים על נחשת ויהיה מועתק לכסף בע״ה . (נזר נסה זה השער וצלות בתנור הלחם חם מעט ובפעם הו׳׳ נתך)

[§ 55] **רציאו** בבדיל שהוא אלוקי ואלקלעי. והוא מחלק צדק וטבעו חם ולח

fol. 32v

you want, cut them with scissors and put them in a *crucible*[448] with some *borax*[449] or *natron*,[450] which is the salty stone. Then, when it will be dissolved *in my opinion: melted*, put one *dirham* of it by itself – in my opinion: of iron or copper – with one *dirham* of white *elixir* and add one *dirham* of silver. It comes out white silver that melts like silver, resists fire like silver, withstands the hammer and does not change, with the help of God.

[§ 54. Making silver with copper and mercury]
Another fine and good [preparation]. Take one *pound* of copper *filings* and the same amount of mercury – in my opinion: tin – another interpretation: three times the amount [of copper] – and grind it well in an *antribulo*[451] until it becomes like a dough. After that, wash it well with water and vinegar, until it is completely clean. Then wash it with clean water and let it dry. Take one *pound* of *atrament*, dissolve it in double its quantity of aged urine of young boys and filter [it]. Put on it one *pound* of *sal ammoniac*[452] – in my opinion: watering and roasting. *Cerate* it little by little with this water until it drinks all of it: it will be dyed with a colour between yellow and red, which is called in the Arabic language *aṣfar*.[453] (In my opinion: smear with it sheets of clean silver or of silver or of tin, warm them up little by little until it penetrates them. Do this until it is gold. Then melt it and put it on them). We have already combined mercury and copper. Then use it and rejoice – in my opinion: or roasting, congealing in a full glass sphere, then water it until it congeals and does not smoke. Otherwise, melting and congealing until it is ready, or distilling or congealing in a crucible covered with another crucible, according to Ǧābir's method for the red and the yellow dye. Or filling a *vial* and coagulating on a fire of bran or in hot ash. Otherwise, congeal and melt. And if you want to whiten, put some Yemenite *alum* instead of the *vitriol*,[454] and work with it as we said, that is, put it on copper and it will solidify as silver, with the help of God. (Nazar[455] tried this chapter –[456] roasting in a slightly hot bread oven. The sixth time, it melted).

[§ 55. On tin/*qalaʿī*]
Method for tin, which can be *al-ūqī*[457] or *al-qalaʿī*.[458] It is from the part of Jupiter and its nature is hot and humid,

[448] In transliteration from the Arabic بوط .
[449] In transliteration from the Arabic تنكار .
[450] In transliteration from the Arabic نطروب .
[451] See note 446, above.
[452] In transliteration from the Arabic نشادر .
[453] In transliteration from the Arabic أصفر .
[454] In transliteration from the Arabic زاج in the text.
[455] On the identification of this character see *Introduction*, § 5.7.
[456] The text of the manuscript presents a *lacuna* here, possibly caused by damage in its *Vorlage*.
[457] In transliteration from the Arabic أوقي .
[458] In transliteration from the Arabic قلعي .

fol. 33r

אבל חומו פחות כי הקומפליסיאוני שלו הוא קורוטה כי השולפו נצח אותו במנירא ובו הוא רכות וחולי ומהירות התכה והוא חמוץ בטעמו. א״ל מי שירפא הארבעה חלאים האלו ירפא יחזור ככסף ב״ה. ומקבל הצבע ויעמוד וילבין הנחשת מסגולה כי הוא דומה לו ויקבל צבע אדמימות ויעשה ממנו שמש טוב. ס״א מלח. ויעשה ממנו או עמו מים חדים שיכבשו עמו הכוכב. והוא שונא לכסף ולנחשת והוא ראוי לעשות ממנו כסף וזהב.

[§ 56] **לפמא** דלקונדימנטו שלו הוא שתשים ממנו כרצונך בכלי ברזל הנקרא בטוליטא קקולא. או בתרווד ברזל ותתיכנו ושים עליו כמשקלו מלח כתוש . ובלבל תמיד במטה ברזל עד יעשה אבק. אחר תוציאנו וטחון אותו היטב ושים במזוכך וסתום הפרק וטח מטיט החכמה ובתנור הלחם לילה א״ שים ס״א ובאש חזק. ותקרר ותוציאנו וטחון והרתח אותו במים די שיכסהו והנח עד שירד לתחתית הכלי אחר תוציא המים המלוחים ושים עוד מים חם ותבלבל היטב והנח שירד למטה אחר סנן. כן תעשה עד שיהיה המים היוצא ממנו מתוק בלא טעם מלח ויהיה לבן. אחר תייבש ויעשה סיד. אחר תקח כמוהו אלומי יימיני וטחון והתירנו בכפלים כמוהו חומץ חד. או בשתן ישן. ושים עליו כמוהו נשדרא מצרי לבן. והנח בו עד שיותר ואנצירא סיד הבדיל למעט מעט טחון והשקות ושים ליבש בשמש וכשיתיבש טחון אותו והשקה לו טיפה טיפה ולא תשקהו כל המים יחד. כן עשה עמו למעט מעט עד שיהיה אנצירטו וירכך. ותראהו ב״י או ב״א פעמים שירפק בידיך ובפורפירא ויהיה עשוי ממנו חלקי״ יותר רכים וחֲלָקִים נ״ל שתתירנו אז וישמע הכוכב ויכבשנו. תקח מן הנחשת א״ אונקי ותתיך באלביט. וקח מן הרפואה א״ דרמי ושים על הנחשת המנוקה. ותנפח היטב במפוחים עד שיותך ויהיה כסף טוב ב״ה.

fol. 33r

but its hotness is little, since its *complexion* is *corrupted,* because *sulphur* defeats it in the *mine*. It has softness, illness, quickness of melting and it is sour to the taste. But he who heals these four illnesses, heals it and makes it return like silver, with the help of God. It receives the dye and it is stable; it whitens copper for its quality, because it is similar to it, and receives the red dye. From it the good Sun is made – another interpretation: salt. From it or with it a sharp water is produced with which they subdued mercury. It is an enemy for silver and copper and it is suitable for making silver and gold.

[§ 56. Making silver from tin/*qalaʿī* (i)]

The knowledge of its *preparation* is that you put as much as you want of it in an iron vessel that in *Toledo* is called *qqul'*[459] or in an iron ladle and melt it. Put over it the same weight of pulverised salt and keep on stirring with an iron ladle until it becomes dust. After that, take it out, grind it well and put it in a glass container. Seal the joint, cover it with the clay of wisdom and put it in a bread oven for one night – another interpretation: and in strong fire. Let it cool down and take it out. Grind it and boil it in enough water to cover it. Let it rest until it descends to the bottom of the vessel. Take out the salty water, add again hot water, stirring well. Let it rest so that it descends to the bottom and filter. Do this until the water that comes out from below does not have a salty taste: it is white. After that, let it dry and form a calx. After that, take the same quantity of Yemenite *alum*, grind it, and dissolve it in double its amount of sharp vinegar or in aged urine. Put on it the same amount of white Egyptian *sal ammoniac*[460] and let it rest in it until it is dissolved. *Cerate* the calx of tin little by little by grinding and watering and place it in the sun to dry. When it is dried, grind it and water it drop by drop; do not pour all the water at once. Do this with it little by little until it is cerated and has softened. At the twelfth or eleventh reiteration, you will see that it sticks to your hands and on *porphyry*, and some of its parts will be made softer and smoother – in my opinion: that it has dissolved – and then you will hear the mercury which will be subdued. Take one *ūqiyya*[461] of copper and melt it in a *crucible*.[462] Take one *dirham* of the medicine and put it on the cleaned copper. Blow well [on the fire] with bellows until it melts: it is good silver, with the help of God.

[459] I have not been able to identify the name of this iron vessel, which is here transliterated in Hebrew letters.
[460] In transliteration from Arabic نشادر .
[461] In transliteration from the Arabic أوقية .
[462] In transliteration from Arabic بوط .

fol. 33v

[§ 57] ד״א טוב מזה. קח חומץ חד ותתיר בו מלח הלחם ונשדרא. לכשיותר בו אז תשמור. אז קח בדיל כרצונך ותתיך בתרווד ברזל ואז שים עליו רסינא והניחהו שם עד שישרף ואז תכבהו במי המלחים הנז״ למעלה כן עשה פעמים עד שיהיה עשוי בטוב. כי זה הוא נקויו. ואם תרצה שים מזה המים על הבדיל עד שיחסר. ר״ל שיעשה אפר. אחר טחון עמו עד שישתה כל המים המלוחים. אחר תפעול בעת שתרצה ותגיע לתאותך ואשבע באדני כי זה הדבר הוא סוד שכל החכמים סתמוהו. השם יבינך כי הוא להבינתך קל. נ״ל שאם תרצה להתירו תתירנו. ותשקה הרוחות המוסערות עד יזובו ולא יעשנו הגעת. הלא תראה כי אלו הם חלב הבתולה.

[§ 58] **רציאו** בעפרת אסרוב. והוא קר ויבש. והוא מחלק שבתי. ובו הוא זהב וכסף איפינציאלי מינטי. לא בראיית העין. והוא כבר בגופו אי ריטרדו מובימינטו. והוא מקבל הצבע. וכשיקבל הצבע לא יפסידנו לעולם. וממנו יצא כסף. נ״ל וגם יצא ממנו זהב. ויעשו ממנו מרתק וצירוסי. ואסדנג. ויהיה ממנו אקציר גדול לאדום. וללבן. וממנו גם כן יעשו מימות שכובשין הכוכב. והמים שלו ראוי לברזל. ובאקציר אומר מן הברזל מי צירוסי מתיר הברזל. נ״ל גם אם תכבה בם הברזל יתרכך ויותר ובעת שיתערב עמו לא יפרד ממנו לעולם. ואם יתערב עם הבדיל לא יפרד ממנו. ויהיה מליורטו עם נחשת קונדיטו. ובטוב קונקורדא יתחבר עם הכסף. אך יפרד ממנו עם דיקוזיאוני לוקוולי סולי פורִיפִּיקָרֵי הכסף. אבל הוא לא יהיה אסוצטו לזהב. ולא עמו סימלייורטו ולא ישמח אליו ריחו יפסיד ויפרד הזהב וריחו יקפיא הכוכב. ופלוסופי הינדי גם כן אמרו כי הוא יותר קרוב ושכן לזהב מכל הגופות. כי נסתר הזהב הוא קר ויבש. ונסתר האסרוב הוא חם ולח. ובהפך כי נגלה הזהב חם ולח ונגלה האסרוב קר ויבש. א״ך גלוי הוא כי נגלה האסרוב הוא קר ויבש

fol. 33v

[§ 57. Making silver from tin/*qalaʿī* (ii)]

Another better than this. Take sharp vinegar and dissolve bread salt and *sal ammoniac*[463] in it. When they are dissolved in it, put it aside. Then take as much as you like of tin and melt it in an iron ladle. Then put some *resin* on it and let it rest there until it burns. Then quench it in the aforementioned water of salts. Do this multiple times until it is well made, because this is its purification. If you want, put some of this water on tin until it is absorbed – this means: that makes ash. After that, grind [with] it until it drinks all the water of salts. After that, use it when you want and you will obtain what you desire. I swear by my Lord that this thing is the secret that all the sages hid. May God let you understand [it] because it is easy for your understanding – in my opinion: that if you want to dissolve it, then dissolve it. Water the sublimed spirits until they drip, they do not smoke and [produce] sound. Don't you see that these are the milk of the virgin?[464]

[§ 58. On lead]

Method of lead *usrub*.[465] It is cold and dry and is from the part of Saturn. *Essentially*, there are gold and silver in it, but do not appear to the eye. It is heavy in its body and *retarded in its movement*. It receives the dye and, when it has received the dye, it does not leave it forever. Silver comes out of it – in my opinion: also gold comes out of it. From lead [the alchemists] made *litharge*,[466] *ceruse* and *white lead*.[467] From it comes the great *elixir*[468] for the red and for the white and from it they also made waters that subdue mercury. Its water is suitable for iron. With regard to the *elixir*, it is said that from iron [comes] a water of *ceruse* that melts iron – in my opinion: also if you quench iron in it, it softens and it melts. When you mix [something] with lead, it will not separate from it forever. If it is mixed with tin, it does not separate from it. It is *rectified* with *prepared* copper; it *agrees* well and mixes with silver and separates from it only through the *decoction that is employed for purifying* silver. Anyway, it is not *associated* with gold that does not *become better* with it nor does it rejoice in its presence. The exhalations of lead damage gold and make it divide. Its exhalations congeal mercury. Indian philosophers also said that lead is closer and nearer to gold than all other bodies, because the interior of gold is cold and dry, while the interior of *lead*[469] its hot and humid; on the other hand, the exterior of gold is hot and humid, while the exterior of *lead* is cold and dry. After that, it is clear that the exterior of *lead* is cold and dry

[463] In transliteration from Arabic نشادر in the text.

[464] "The milk of the virgin" is a common *Deckname* in alchemical literature, that can indicate ingredients that enter into an alchemical preparation, but also the final product or the elixir.

[465] The manuscript has here two terms for lead: the Hebrew name of the metal עפרת is followed by the Arabic name أسرب in transliteration.

[466] In transliteration from Arabic مرتك in the text.

[467] In transliteration from Arabic إسفيداج in the text.

[468] In transliteration from Arabic إكسير in all occurrences in this section.

[469] In transliteration from Arabic أسرب in all occurrences in this section.

fol. 34r

נסתרו חם ולח. ובהפך הבדיל קלעי. כי הקלעי נסתרו כסף. ונגלהו זהב. כי נמצא כסף וזהב יותר קרובים ושכנים לאסרוב מן הקלעי ויותר יהפך לזהב וכסף. והוא מיותר דמים ומרבוי גדול. אמר לי נזרי כי נסה והדליף טסי בדיל ויצא ממנו מים לבן ויליני. ומטסי נחשת יצאו מים אדומים. ואחר יצאו מים שנטו לשחרות

ואמר פתאגורש כי העפרת הוא גדולה. וכל הסוד בו סתום. והוא אשה רכה. ואמר כי האשה הזאת יש לה ג׳׳ עניינים. והוא לובן אודם שחרות. ועוד ד׳׳ דברים אחרים. רכוך. והתוך בקלות. ויובש כי היא שולפורטו. ושורפת הקרירות. נ״ל ומקררת השריפה כי היא תכבה חמימות הזכר שלה נ״ל כי היא מכבה חמימות הגפרית. כי הן שמו לה זה השם מי הגפרית. ואין שום דבר יותר יכולה ויותר שכנה ויותר חזקה הכרנו ממנה הכרנו מאותה אשה.

ואמר אסידירו אל תאמרו כי האסרוב הוא מעופרת הקלעי ומגופות אחרות. אבל הוא זהב אשר באותו נכנסו חלאים בו במנירא והעתיקוהו. כמו שהחולי יכנס בוולד במעי אמו. והוא על הזהב בכל ענייניו. והזהב הוא גוף קיים וכבד אִידְימוטֵטו. והוא יעמוד בארץ ויעמוד באויר ובמים ולא יפסד. והוא כמו שאומרים. א״כ אז כשתרצה לנסות ולהבחין האמת מן השקר בעפרת. קח כלי חרס ושרוף אותו. ר״ל העופרת עם אש טרבירצו כמו שעושים היוצרים בקערות מזוככות אז אותו יעשה צירוסי ואם עדיין תדליק בו האש אותו יעשה אַזְרנוֹך

[§ 59] **לפמא** דילוסואו קונדימינטו הוא שתקח ממנו חלק ותתיך בתרווד ברזל ושים עליו כמשקלו כוכב. ושים בפורפירא וקח כמשקל אחד מהם מלח צלוי ותתירנו בחומץ חד כפלים כמוהו ואחר סנן

fol. 34r

and its interior is hot and humid. Tin *qalaʿī*[470] is the opposite, because the exterior of *qalaʿī* is silver and its interior is gold, since we have discovered that silver and gold are closer and nearer to *lead*[471] than to *qalaʿī*,[472] that [lead] becomes more easily gold and silver, and that it is very similar to them and is a great multiplicator. Nazari[473] told me that he tried to distill sheets of tin and a white [and] *yellow* water came out of them and that a red water came out of sheets of copper and that after that a water that tended to black came out. Pythagoras **said** that lead is great[474] and that every secret is hidden in it. It is the gentle woman. He said that this woman has three matters, which are white, red and black. And she also has three other things: softness, ease of melting, dryness because it is *sulphureous*, and it is a refresher of combustion – in my opinion: it cools down combustion because it extinguishes the male hotness in it – in my opinion: because it extinguishes the hotness of sulphur. For this reason, they gave it this name: water of sulphur. We have never known anything more powerful, closer and stronger than what we learnt from this woman. *Isidoro* **said**: do not say that *usrub* derives from lead *qalaʿī* or from other bodies. It is instead gold in which illnesses penetrated in the *mine* and transformed it, like illnesses enter the fetus in his mother's womb. It is [related to] gold in all its features. Gold is a stable, heavy and *silent* body; it perdures on the earth, perdures in the air and in the water without degrading. It is as they said. And then, if you want to try and distinguish true and false in regard to lead, take a clay vessel and burn it – this means: lead – with *transverse* fire like potters do with glazed bowls. Then [lead] becomes *ceruse* and, if you light again the fire on it, it turns it into *red lead.*[475]

[§ 59. Preparation of lead (i)]

The Knowledge *of its preparation* is that you take one part of it and you melt it in a iron ladle. Put on it the same weight of mercury and put it in *porphyry.* Take of roasted salt the weight of one of them and dissolve it in twice its amount of sharp vinegar. Then filter

470 In transliteration from the Arabic قلعي in all occurrences in this section.
471 In transliteration from the Arabic أسرب in all occurrences in this section.
472 In transliteration from the Arabic قلعي in the text.
473 On the identification of this authority, see *Introduction*, § 5.7.
474 The adjective is here in the feminine form; this could be interpreted as a mistake of the translator/copyist or an intentional indication of the feminine nature of lead, which is mentioned in the following lines.
475 In transliteration from the Arabic أسرنج .

fol. 34v

ותשקם ר״ל העופרת למעט מעט טחון והשקות ויבש עד שישחיר. אחר תשרוף כראוי. נ״ל בתנור צלה א״ יום וא״ לילה עד יעשה לבן אחר תרחצנו מן המלח עד ימתיך ועשה עמו ב״ה. ואם תרצה שים עמו נשדרא והשקה לו אותם ויהיה טוב למאד. נ״ל תתיר בחומץ ותשקה כוכב מוסער וזרניך מוסער עד יזובו כשעוה. או תקפיא עם המים ההם הכוכב.

[§ 60] **ד״א** טוב מאד. קח אסרוב או אלוק קלעי איזה שתרצה ותתיך בתרווד ברזל ותטעים בכל ליטרא מהם א״ אונקי מגפרית נ״ל לאדום. או א״ אונקיי מזרניך נ״ל ללבן. ובלבל אותו עם מטה עד שיהיה כאפר שחור א״ שעה פתאום. אחר תטחון אותו עם כמו *נ״ל כמוהו* מלח צלוי מותר בחומץ חד. וכשישתהו כלו שים בקדרה מזוככת שיסבול האש והדלק תחתיו האש י״ שעות עד שיצא סיד לבן.

[§ 61] **ד״א** תתיר מלח בכפלים חומץ חד. וקח לימאטורא מאסרוב או לימטורא מאלוק. וטחון למעט מעט עמו עד שישתהו כלו. אחר תשים על האש כאשר אמרנו.

[§ 62] **אלביאד** לבן הוא צירוסי של עופרת. קח טסי עופרת כרצונך ותלה בכלי מלא חומץ שיהיה בו ענבים חמוצים. ר״ל בוסר. וכסה אותם עד י״ ימים ותמצאם הגרב הלבן גרר אותו ושים בכלי חרס עד ד״ ימים והנח עד שיצליל. והוצא המים בנחת וקח השמרים ושים במקום שבעלי זכוכית קוראים[a] מוֹיְרְקוֹן. ולוקחין אותו שנכבר. ושים בכלי עם מים ומלח מעט ובלבל ושפשף היטב ביד. והנח לנוח. ויעלה המים. וסנן. ושים מים אחר עד שיעשה חם. אחר שים אותו בלבינה או באבן לשמש. אז יבא הרבה טוב והרבה לבן נעשה וגם כן אלבייד מלימטורא של עופרת כמו שהמיניאו נעשה מלימטורא של נחשת

---

[a] קורים

fol. 34v

and water – this means: the lead – little by little, by grinding and watering, and it will dry until it darkens. After that, burn it as appropriate – in my opinion: roast it in an oven for one day and one night until it is made white. Then wash it from the salt until it becomes sweet and use it, with the help of God. And if you want, put *sal ammoniac*[476] with it and water it with them: it will be very good – in my opinion: dissolve it in vinegar and water [it with] sublimed mercury and sublimed *arsenic*[477] until they drip like wax or congeal mercury with these waters.

[§ 60. Preparation of lead (ii)]
Another very good. Take the one you prefer between *lead*[478] or *tin qalaʿī*[479] and melt it in an iron ladle. Feed every *pound* of it with one *ūqiyya*[480] of sulphur – in my opinion: for red – or one *ūqiyya* of *arsenic* – in my opinion: for white. Stir it with a stick until in one hour it looks suddenly like dark ash. After that, grind it in like *in my opinion: the same amount* of roasted salt dissolved in sharp vinegar and, when it has absorbed it all, put it in a glass pan that withstands fire and light a fire under it for ten hours until white calx comes out.

[§ 61. Preparation of lead (iii)]
Another. Dissolve salt in double its amount of sharp vinegar. Take *lead filings* or *tin*[481] *filings* and grind them with it little by little until it absorbs it all. Then put it on a fire like we said.

[§ 62. On the white egg]
The white egg is the *ceruse* of lead. Take as many sheets of lead as you want and hang them in a vessel filled with vinegar in which there are sour grapes – this means: unripe. Cover them for ten days and you will find the sock white.[482] Take it out, put it in a clay vessel for three days and let it rest until it sinks. Take out the water from the bottom, take the dross, put them in the place that the glass-makers call *moyerqon*,[483] and take what has been buried. Put it in a vessel with water and a little salt, mix it, and rub it well with the hand. Let it rest: the water will rise. Filter and put on it other water until it becomes hot. Then put it in the sun on a brick or on a stone. Then it will become very good and very white. In this way the egg white from *filings* of lead is also made, like *red lead* is made from *filings of* copper.

476 In transliteration from the Arabic نشادر .
477 In transliteration from the Arabic زرنيخ in all occurrences in this section.
478 In transliteration from the Arabic أسرب .
479 In transliteration from the Arabic آنك قلعي .
480 In transliteration from the Arabic أوقية in all occurrences in this section.
481 In transliteration from the Arabic آنك in the text.
482 The sock is probably a cloth envelope in which lead is wrapped and then suspended in the vase of vinegar, although the recipe does not mention it before this passage.
483 I have not been able to identify the kind of glassmaking apparatus employed here.

fol. 35r

[§ 63] **רציאו** של זכוכית והוא מחלק כוכב. לא אדע למה יהיה זה אלא בעבור שמקבל הצבע והדמות. וגם הוא יתיר הברזל וכל הגופות ויעשה אותם דיקוֹרֵירֵי אִי דישוֹלֵיירֵי. אבל אני באלו לא נסיתי אלא באבני של טבעות וקצתן אומר ב״ה

[§ 64] **מלאכת** רובינו. קח זכוכית טוב הבא דְלַאִירַק. ותתיך באלביט וכשיותך שים עליו שדנא קונדיטא ובלעז אמיטיטים ואחר תשימו באיזה דפוס שתרצה ותתיך עם מעט מלח אלקלי. וכשיותך שים בו משקל א״ גרגיר מקרוקו של ברזל קונדיטו ממי אטרמנטו. ואם ייטב בעיניך דמותו הרי מה טוב ואם לא הוסף מן הרפואה יותר ויהיה פלידא ודומה ליקווינטו ב״ה

[§ 65] **מלאכת** יימי רוביאי. תתיך מזכוכית הנז״ כמו שעשית למעלה ושים משקל א״ גרגיר ממגניסיאה מלובנת קונדיטא ועם אבן אזוורד. או מותר במי אלומי ונסה למעט מעט. ואם יהיה דמותו בעיניך (כענן) כתו״ טוב יהיה. ואם לא עשה עד שיהיה קולורטו ב״ה

[§ 66] **מלאכת** יימי ירוקות. תתיך כאשר אמר״ ושים על משקל גרה זינגייר ומעט מותר עד שייטב לך ב״ה

כבר השלמנו המאמרים מן הגופות והרוחות והמלחים חדין קונדיתו. ונשאר שנאמר מה שנשאר

[§ 67] **נתחיל** מהתשיעי דקונדימנטו של אבנים. וההרכבות הקלות אותו שהוא הרבה צריך לתלמידים ולעמים. והעמים מלאכות גדולות מרכיבים. שער הקונדימנטו מהתשיעית. קח מן התשיעית לבן כרצונך וטחון עד שכולו יתדקדק. אחר קשור בבגד פשתים נקי ומהודק וקח קדרה מזוככת ושים בה פולין יבשים די צרכך וכהם ממים ושים על האש. ותלה הבגד אשר בו התשיעי. ~~אלו כלו יהיה מותך בקערה~~ באיד הפולין עד שתראהו נכנס מקובץ והנח א״ שער. ~~ואחר תוציא וקח קערה מזככת~~

fol. 35r

[§ 63. On glass]
Method for glass. It is from the part of Mercury. I do not know the reason for this, except from the fact that it receives the dye and its appearance. It also melts iron and all the bodies and it makes them *run and come up*. But, among these things, I have not tried anything apart from the stones for rings. I will talk about them briefly, with the help of God.

[§ 64. Preparation of artificial ruby]
Preparation of *ruby*. Take good glass that comes *from Iraq,* melt it in a *crucible*[484] and, when it is melted, put over it *prepared* amethyst, which is called in foreign language *amethyst*. After that, put it in any mould you want and melt it with a little *alkali*[485] salt. When it is melted, put on it the weight of one grain of *crocus* of iron prepared with water of *atrament*. If its colour looks good to you, then this is a good thing; if it does not, add more medicine: it will be *pale* and similar to *hyacinth,* with the help of God.

[§ 65. Preparation of an artificial azure gemstone]
Preparation *of red gems.*[486] Melt some of the aforementioned glass as you did before and put the weight of one grain of whitened *prepared magnesia* with one stone of *lapis lazuli*[487] or dissolved in water of *alum*. Try little by little. If its colour looks azure to you (like a cloud), this is good; if not, continue until it becomes *coloured*, with the help of God.

[§ 66. Preparation of an artificial green gemstone]
Preparation of green *gems*. Melt like we said, put on it the weight of a grain of *verdigris*[488] and melt it slowly until it is fine for you, with the help of God.

We have now completed the discourses on bodies, spirits and the sharp and *prepared* salts. It remains to talk about what remains.

[§ 67. On the ninth and its preparation]
We begin with the ninth[489] for the *preparation* of stones and with its simple procedures, because it is very necessary for the students, the people, and those who compose great works. Chapter[490] on the *preparation* of the ninth. Take as much white ninth as you want and grind it until it is all crumbled. After that, tie it in a clean and tight linen cloth and take a glass pan. Put as many dried fava beans in it as you need and the same amount of water and put it on the fire. Hang [on it] the cloth in which there is the ninth.[491] with the exhalation of the fava beans, until you see that it starts to form lumps. Let it rest for an hour. ~~Then take it out and take a glazed bowl~~[492]

[484] In transliteration from the Arabic بوط .
[485] In transliteration from the Arabic القلي .
[486] The title of this recipe does not match its content: while the recipe describes the preparation of an azure gem, the title promises a recipe for a red gem.
[487] In transliteration from the Arabic لازورد .
[488] In transliteration from the Arabic زنجار .
[489] The substance here indicated with the progressive numeral "ninth" has been identified by Patai, *Jewish Alchemists*, 122–3 with talc. Patai maintains with some reason that the seven metals used by alchemists could be called by their progressive number from one to seven and, therefore, argues that the "ninth" should be identified with talc. Patai does not provide any supporting evidence to this claim. I have therefore preferred to keep the numeral in my translation.
[490] The term שער meaning "gate" and here indicating a new "chapter" mirrors the Arabic usage of the term باب , literally "door."
[491] This portion of text is crossed out in the manuscript.
[492] This portion of text is crossed out in the manuscript.

fol. 35v

~~ושים בה פולין יבשים די צורכך וכסה ממים ושים על האש. ותלה הבגד אשר בו התשיעי באיר הפולין עד שתראהו נכנס מקובץ וסנח א״ שער.~~ ואחר תוציא וקח קערה מזוככת ותתיר הקשר ושים בו גויגֶם כרצונך. ושפשף היטב בידיך עד שלא ישאר בבגד מן התשיעי אלא כולו יהיה מותר בקערה כמו כסף חי. ואם תשים בקערה מעט מי פולין עד שהבגד יטבע בו והוא ר״ל התשיעי יהיה מותר אז ירד בתחתית הקערה כמו כסף חי. אז כשתראהו סנן המים ממנו למעט מעט עד שתראהו זך. אחר תיבשהו ושמור. כי הוא כמו הקלס. אז תקח ממנו חלק. נ״ל או תקח מהרביעית או מחלזון. או מפירני. וכמוהו כוכב מוסער ומוקפא. וכמותן נשדרא מותר. ר״ל כתשיעית וככוכב וטחון אותו עם מי הנשדרא עד שיתערבו היטב. אחר תקבץ ושים במזוכך. וסתום היטב הפרק ושים בארץ במקום לח מ״ יום אז תמצאנו מותר. אז סבב המזוכך אפר והדלק אש רפה עד שיקפא כחלב וכספון. אז שים א״ חלק על מאתים קלעי נקי ויועתק לכסף ב״ה . ונ״ל כי גם הקלעי יקפיא.

[§ 68] **שער** אחר טוב ויפה. קח מלח אלקלי כתוש ותתיר בבודילו. וקח מתשיעי כרצונך. מאותו שהתרת במי הפולין. וכמוהו זרניך מלובן מוסער. וטחון עם זה המים ויבש עם מלח הרבה פעמים. או על אפר חם עד שיעשה כחלב ברכות ובלחלות. שים במזוכך עד שיותר כשעוה ועשה עמו במה שתרצה. נ״ל בבדיל. ובכוכב מוקפא.

[§ 69] **שער** אקונדימנטו דמרקשיתא. היא אבן מירקוליאלי. והיא כספית וזהבי״ ולא נסיתי רק אלו הב״ מינים. ומצאתי בספר הסגולות כי מרקשיתא יתיר הברזל כשתזרה עליו. וישרוף הנחשת והכסף. וישבר הזהב ועוצר עפרת קלעי. [§ 70] לפמא דילו סואו קונדימינטו הוא שתקח ממנו כרצונך. (נ״ל שרוף בטוטיאה וכבה ז״פ בחומץ ויבש והשקה מי מלח

fol. 35v

~~and put in it as many dried fava beans as you want. Cover them with water and put it on fire. Hang the cloth in which there is the ninth on the exhalations of the fava beans until you see that it starts to form lumps. Let it rest for an hour.~~[493] Then take it out, take a glazed bowl, and undo the knot. Put in the bowl as much *gwyygm*[494] as you want. Rub it well with your hands until no ninth remains in the cloth, since it has all melted in the bowl like quicksilver. And if you put in the bowl some water of fava beans until the cloth sinks in it and it – this means: the ninth – is dissolved, then it will deposit at the bottom of the bowl like quicksilver. Then, when you see it, filter the water from it little by little until you will see that it is pure. After that, let it dry and save it because it is like lime. Then take a part of it – in my opinion: or take some of the fourth or of snail or of *plums* –, the same amount of sublimed and congealed mercury and the same amount of the two ingredients of dissolved *sal ammoniac*[495] – this means: like the ninth and mercury – and grind it with the water of *sal ammoniac* until they are well mixed. Then collect them and put them in a glass vessel. Seal the joint well and put the vessel in the ground in a humid place for forty days. Then you will find them dissolved. Then cover all around the vessel with ash and light a gentle fire until it congeals like milk and *soap*. Then put one part on two hundred parts of pure *qalaʿī*[496] and it will become silver, with the help of God – in my opinion: because the *qalaʿī* congeals too.

[§ 68. Preparation of the ninth (ii)]
Another **chapter** good and fine. Take pulverised *alkali*[497] salt and dissolve it in a *piece of gut*. Take as much as you want of the ninth that you previously dissolved in water of fava beans, and the same amount of whitened and sublimed *arsenic*.[498] Grind it with that water and dry it many times with salt or on hot ashes, until it becomes like milk for softness and humidity. Put it in a glass vessel until it melts like wax and use it with anything you want – in my opinion: with tin and with congealed mercury.

[§ 69. On marcasite]
Chapter. The preparation of marcasite. It is a *mercurial* stone and it is silvery and golden. I did not try anything apart from these two kinds. I have found in the *Book of Qualities*[499] that marcasite melts iron, when it is thrown on it, burns copper and silver, breaks gold and restrains *qalaʿī* lead.

[§ 70. Preparation of marcasite]
The knowledge of its preparation is that you take as much as you want of it (in my opinion: burn it with *tutty* and quench it – this is a commentary – seven times in vinegar. Let it dry and water it with water of salt

493 The copyist, having recognised that this section was a duplicate of the one in the previous folio, crossed it out in the manuscript.
494 I have not been able to identify this ingredient, whose name appears here in transliteration from an Italian word.
495 In transliteration from the Arabic نشادر in all occurrences in this section.
496 In transliteration from the Arabic قلعي in all occurrences in this section.
497 In transliteration from the Arabic القلي .
498 In transliteration from the Arabic زرنيخ .
499 The very general title of this work prevents a clear identification of authorship and time of production.

fol. 36r

אלקלי עד יותך וכו״ל). וכתוש וטחון היטב אֶיבוֹלבִּירַאי נ״ל טחון אותו סל צאבון. נ״ל שמן ובורק. או ספוני ישמעאלית ועם זכוכית. נ״ל נטרון. ותתיך נ״ל ותוריד באלביט. נ״ל ברבוט. וכשיותך תוציא השמרים. ואם תעשה זה ב״ וג״ פעמי״ יהיה יותר טוב. אז תקח הזך שבו. והוא הנקרא בלשון ערבי כוקרא. ונעשה פרוסטו לבן דומה הרבה לכסף אך כי ישבר תחת הפתיש.

ואם תרצה לטוחנו. יטחנו. ותתירנו. (ס״א תשקנו) בנשדרא מותר בחומץ חד נ״ל וגם אלומי ללבן. ולכרכומית עם זאג ואנצירא עמו אותו עד שישתהו כולו נ״ל כמוהו. אז יהיה הוא אקציר יפה הרבה וטוב ולא תצטרך מאחר ב״א ושים ממנו על א״ מבדיל מנוקה ויצרנו וישקנו. ויקח קולו וריחו. נ״ל כי זה יקפיא הכוכב. ואחר תשליך ממנו א״ על נחשת אדום וילבינו ליבון נאה. אז תחברהו עם כמוהו כסף. ושניהם יהיו לך אם תרצה. ויהיה לך תועלת ביום אחד ויותר טוב ויותר קרוב ויותר תועלת שיוכל להיות. ודי לך פרי טוטי לאלטרי אם כן תסתיר.

fol. 36r

alkali[500] until it is melted with heart and mind).[501] Pulverise it, grind it well, and *stir it* – in my opinion: grind that salt of soap – in my opinion: oil and *borax*[502] – or Ismaili *soap* and with glass – in my opinion: *natron*. And melt it – in my opinion: let it descend – in the *crucible*[503] – in my opinion: *double crucible*.[504] When it melts, take out the dross. And if you do this two or three times, it will be better. Then take the pure part that is in it, and this is what in the Arabic language is called *kwqr'*.[505] It *shortly* becomes white [and] very similar to silver, with the exception that it will break under the hammer. If you want to grind it, grind it and then melt it (another interpretation: water it) with *sal ammoniac*[506] dissolved in sharp vinegar – in my opinion: and also alum – for the white, with *vitriol*[507] for the yellow. Cerate it with this until it has absorbed it all – in my opinion: like its quantity. Then this is the very fine and good *elixir*[508] and you will not need anything else, with the help of God. Put some of it on one part of purified tin: it will subdue it, water it and it will take away its voice and its smell – in my opinion: because this coagulates mercury. Then throw one part of it on red copper: it will whiten with a pure white colour. Then unite it with the same amount of silver. They are both going to be [good] for you, if you want. Here you have something useful in one single day and it is better, closer and more useful than it can be. May this be enough for you *for all the others*. If so, hide it.

[500] In transliteration from the Arabic القلي .

[501] I am proposing a very provisional translation of this abbreviation, which could be expanded as כליות ולב, meaning "kidneys and heart" and, therefore, "emotion and reason." The phrase is Biblical (Jer. 11,20; Ps. 7,10) and it is also found in the Babylonian Talmud (Beraḥot 61).

[502] In transliteration from the Arabic بورق .

[503] In transliteration from the Arabic بوط .

[504] In transliteration from the Arabic بر بوط . The double crucible (known in Latin with a rendition of the Arabic name as *botum barbatum*) is an apparatus used for separating dross from pure materials during fusion.

[505] The transliteration of this Arabic word into Hebrew prevents its understanding.

[506] In transliteration from the Arabic نشادر .

[507] In transliteration from the Arabic زاج .

[508] In transliteration from the Arabic إكسير .

APPENDICES

Appendix 1. Table of Contents of MS Sprenger 1908, Staatsbibliothek zu Berlin – Preussischer Kulturbesitz, Orientabteilung[509]

(1) preparation of the *argilla philosophorum*; use of the aludel (fol. 2 *recto*)
(2) generation of animal life from vegetable life; the seven metals (fol. 2 *verso*)
(3) alchemical lexicon of *Decknamen* for metals (fols. 3 *recto*-4 *recto*)
(4) alchemical lexicon of *Decknamen* for volatile substances (fols. 4 *verso*-5 *verso*)
(5) alchemical lexicon of *Decknamen* for stones (fols. 5 *verso*-6 *recto*)
(6) laboratory equipment (fols. 6 *recto*-7 *verso*)
(7) sublimation (fols. 7 *verso*-8 *recto*)
(8) preparation of red sal ammoniac (fol. 8 *recto*)
(9) calcination and dissolution of salt (fol. 8 *verso*)
(10) reddening of the precious stone (fol. 9 *recto*)
(11) calcination of the Sun; calcination and reddening of tutty (fol. 9 *verso*)
(12) red mercury (fol. 10 *recto*)
(13) calcination of gold; dyed iron; calcination of copper; water of vitriol (fol. 10 *verso*)
(14) distillation of the red colour and of hematite (fol. 11 *recto*)
(15) water of vitriol (fols. 11 *verso*-12 *recto*)
(16) distillation of marcasite; operations on animal substances (fol. 12 *recto*)
(17) useful animal substances for the alchemical work; the seven spiritual things entering the operation; calcination of the Moon; water of sal ammoniac and sulphur (fol. 12 *verso*)
(18) calcination of glass (fol. 12 *verso*)
(19) operations on the Moon; dissolution of mercury (fol. 13 *recto*)
(20) quick dissolution of talc (fol. 13 *verso*)
(21) red dye; operations on white lead (fol. 14 *recto*)
(22) operations on the Sun (fol. 14 *verso*)
(23) dyeing of the Sun (fols. 14 *verso*-15 *recto*)
(24) operations on the Moon; sublimation of musk, camphor, and saffron (fol. 15 *recto*)
(25) operations on camphor; sublimation of rose water; sublimation of wood; congealment of the Moon (fol. 15 *verso*)
(26) operations on the Moon (fol. 16 *recto*)
(27) operations on cinnabar; red sulphur (fol. 16 *verso*)
(28) operations on talc (fol. 17 *recto*)
(29) amalgamation of the Sun (fol. 17 *verso*)

[509] I am here presenting an overview of the topics dealt with in the manuscript. Where rubricated headings appear in the manuscript, I simply provide a translation. Otherwise, I provide a brief summary of the content of the relevant manuscript section.

(30) transmutation of the Moon into the Sun; cleaning of gold (fol. 18 *recto*)
(31) cleaning of the silver; operations on the Sun (fol. 18 *verso*)
(32) Pseudo-Rāzī's *On Alums and Salts* (fols. 19 *recto*-30 *verso*)
(33) increase of the weight of gold with silver; preparation of a solvent for stones and metals (fols. 30 *verso*-31 *recto*)
(34) transmutation of base metals into silver; operations on silver (fol. 31 *recto*)
(35) transmutation of glass into a precious stone (fol. 31 *verso*)
(36) operations on the Sun; fixation of cinnabar; yellowing of cinnabar (fol. 32 *recto*)
(37) preparation of gold with mercury; operations on gold (fol. 32 *verso*)
(38) increase of the weight of gold with iron (fol. 33 *recto*)
(39) discourse on the calcination and the operation on the bodies; calcination of the Sun; calcination of the Moon (fol. 33 *verso*)
(40) calcination of Venus; calcination of lead and of *qala'ī*; calcination of Mars (fol. 34 *recto*)
(41) calcination of common stones; melting of the bodies; dissolution of spirits: arsenic and sulphur (fol. 34 *verso*)
(42) operations on metals; operations on the Sun (fol. 35 *recto*)
(43) sublimation of Mars; operation of the amalgam (fol. 35 *verso*)
(44) softening of the Sun (fol. 36 *recto*)
(45) operations on salts (fol. 36 *verso*)
(46) restraining the fleeing [servant] (fol. 37 *recto*)
(47) operations on the Sun; rectification of the Sun; sublimation of the servant (fol. 37 *verso*)
(48) the rust of *qala'ī*; the milk of the virgin (fol. 38 *recto*)
(49) the white elixir for the rectification of *qala'ī*; distillation of vitriol and Mars (fol. 38 *verso*)
(50) congealment of mercury; melting of glass (fol. 39 *recto*)
(51) operations for obtaining gold and silver (fol. 39 *verso*)
(52) whitening of sulphur; fixation of sal ammoniac; melting of copper filings; operations on silver (fol. 40 *recto*)
(53) melting of iron; recipes for obtaining the Moon (fol. 41 *recto*)
(54) congealment of the servant; operation of Hermes (fols. 41 *verso*-42 *recto*)
(55) cleaning of Venus; operations on the servant (fol. 42 *verso*)
(56) dialogue between mercury and gold (fol. 43 *verso*)[510]
(57) operations on gold; calcination of the Moon; operations on green vitriol (fol. 44 *recto*)
(58) operations on pearl (fol. 44 *verso*)
(59) preparation of the Moon with the servant (fol. 45 *recto*)
(60) congealment of the servant (fol. 45 *verso*)
(61) purification of Venus; softening and whitening of Venus (fol. 46 recto)

[510] The same dialogue is also found in the Hebrew version of the *OAS* (fol. 25 *verso*), within a section of the work that has not been preserved in the Arabic manuscript.

Appendix 2. Table of Contents of the Hebrew MS Orient. Oct. 514, Staatsbibliothek zu Berlin – Preussischer Kulturbesitz, Orientabteilung[511]

(1) operations on silver (fol. 1 *recto*)
(2) preparation of a solvent for metals (fol. 1 *verso*)
(3) operations on salt and alum; washing of the egg; preparation of vitriol from glass; preparation of sharp water (fol. 2 *recto*)
(4) congealment of common salt; quotation of Bernard[512] (fol. 2 *verso*)
(5) cooking of lead; preparation of sharp waters (fol. 3 *recto*)
(6) distillation of mercury; preparation of sharp waters; quotation of 'the philosophers' (fol. 3 *verso*)
(7) mixing sulphur with other substances; dissolution of hematite; clay of wisdom; cooking of copper (fol. 4 *recto*)
(8) preparation of waters; operations on silver and iron; congealment of spirits; congealment of mercury (fol. 4 *verso*)
(9) dissolution of borax and alum with moisture; resurrection of sulphur and mercury (fol. 5 *recto*)
(10) whitening of mercury; operation on alkali salt and lime (fol. 5 *verso*)
(11) whitening of sulphur and arsenic; whitening of mercury (fol. 6 *recto*)
(12) whitening of mercury; cooking of arsenic (fols. 6 *recto*-6 *verso*)[513]
(13) congealment of mercury; making sharp water from sulphur; sublimation of iron (fol. 7 *recto*)
(14) sublimation of the two kinds of lead, of marcasite, of gold and of silver (fol. 7 *verso*)
(15) sublimation of copper; dissolution of mercury; dissolution with sharp waters; congealment of lead and iron (fol. 8 *recto*)
(16) difference between kinds of fire; roasting of iron; burning of iron and silver (fol. 8 *verso*)
(17) roasting of iron; rectification of vitriol with water (fol. 9 *recto*)
(18) melting of lead; melting of glass (fol. 9 *verso*)
(19) melting of talc, of glass and of silver (fol. 10 *recto*)
(20) preparation of iron; operation for obtaining the Sun (fol. 10 *verso*)
(21) operation with eggshells and sulphur operation on iron; hematite from iron; quotation of Ǧābir (fol. 11 *recto*)
(22) the truthful operation of *gšpnp*[514] (fols. 11 *verso*-12 *recto*)
(23) operation called 'the Messiah of the righteous' (fol. 12 *verso*)

[511] I here present an overview of the topics dealt with in the manuscript by providing a translation of the headings in quadratic script, when available, or else a brief summary of the content of the relevant manuscript section.

[512] The name is in transliteration from Italian and could be read as Bernardo.

[513] The aforementioned mistake in the foliation in Arabic numerals intervenes at this point of the manuscript; the topics here listed for fol. 6 extend over two folios rather than one.

[514] I have not been able to identify the word, probably a person's name, concealed by this transliteration in Hebrew letters.

(24) melting of arsenic and sulphur; quotation of Mary, the alchemist (fol. 13 *recto*)
(25) quotation of *ryskr*[515]; rectification of mercury (fol. 13 *verso*)
(26) operation with rock salt and Egyptian vitriol; sharp waters; quotation of Ğābir (fol. 14 *recto*)
(27) cleaning of iron; cleaning of the two kinds of lead; congealment of white copper (fols. 14 *verso*-15 *recto*)
(28) description of two substances called 'the root that sprouts and that the philosophers concealed' and 'the gentleman that heals gold and silver by whitening them' (fol. 15 *verso*)
(29) description of a water of alkali salt that whitens and does not burn; quotation of the book 'Treatise on copper' (fol. 16 *recto*)
(30) quotation of Ğābir on the transmutation of silver into gold; operation on red arsenic (fol. 16 *verso*)
(31) on the 'soap of the Sages' (fol. 17 *recto*)
(32) method for cleaning of minerals (fol. 17 *verso*)
(33) description of bitter salt and alum of glass; operation on pure mercury; operations on iron, tutty and alkali salt (fols. 18 *verso*-19 *recto*)
(34) *On Alums and Salts* by Pseudo-Rāzī (fols. 19 *verso*-36 *recto*)
(35) *Chapters of the annotations of the Sages* (fols. 36 *recto*-37 *recto*)
(36) sharp waters for whitening; preparing the 'milk of the virgin' (fol. 37 *verso*)
(37) operations with mercury, lead, arsenic (fol. 38 *recto*)
(38) operations with mercury and tin; operation with mercury and pure lead (fol. 38 *verso*)
(39) operation on borax; preparation of a *margerita* ('daisy') (fol. 39 *recto*)
(40) description of the operations of sublimation and distillation; operation on alum; operation on lime; operation with juices (fol. 39 *verso*)
(41) operations with vegetable juices; operations on copper and sulphur; operation on sheets of silver; operation with sulphur and salt (fol. 40 *recto*)
(42) operation with ground tin and iron filings; operation on silver; operation on tutty; washing of mercury; congealment of mercury (fols. 40 *verso*-41 *verso*)
(43) quotation of Ğābir on the importance of salt in the alchemical work; operations with salt (fol. 42 *recto*)
(44) operations with sharp water (fol. 42 *verso*)
(45) congealment of dissolved substances; operations on crystal; operations on lime; operations with alkali salt; operation with vegetable juices (fol. 43 *recto*)
(46) operations with bread salt and urine; operations with ashes and grease (fol. 43 *verso*)
(47) operations with chicken eggs; operations with atrament and rock salt; operation with sheets of metal; operation with sublimed mercury (fol. 44 *recto*)

[515] The identification of the authority mentioned here is made impossible by the transliteration of their name in Hebrew letters.

(48) ashes found in red alum (fol. 44 *verso*)
(49) operation with arsenic and eggshells; operation with litharge, red arsenic, and alum (fol. 45 *recto*)
(50) operation with sublimed mercury and pure atrament (fols. 45 *verso*-46 *recto*)
(51) the great secret of antimony; mention of Aristotle (fol. 46 *verso*)
(52) congealment of mercury; mention of Arnaldus de Villa Nova; operations for making the Sun (fol. 47 *recto*);
(53) operation for making the Moon (fol. 47 *verso*)
(54) Hermes' medicine; mention of alchemical (and non-strictly alchemical) authorities: Maimonides, Hermes, Ǧābir, Avicenna, Alberto (Magnus), Thomas (Aquinas), (Roger) Bacon; mention of the *Sefer Elim* (fols. 48 *recto* & *verso*)
(55) operations with tutty and mercury; congealment of the servant (fols. 49 *recto*)
(56) rectification of the *zynpry'w* (fol. 49 *verso*)
(57) preparation of red water and of water of lime (fol. 50 *recto*)
(58) whitening of the scorpion; grease of the scorpion (fol. 50 *verso*)
(59) washing of iron filings (fol. 51 *recto*)
(60) grease of the scorpion (fols. 51 *verso*-52 *recto*)
(61) operations on mercury; lime of the Moon (fol. 52 *verso*)
(62) operation with sulphur, mercury, olive oil, and vinegar; operation of the Sun (fol. 53 *recto*)
(63) operations for the production of gold and silver (fol. 53 *verso*)
(64) operation called 'the Messiah'; whitening of the scorpion; congealment of the servant (fol. 54 *recto*)
(65) operation with bitter salt and bread salt; congealment of the servant (fol. 54 *verso*)
(66) operation with the scorpion; sublimation of the eagle (fol. 55 *recto*)
(67) operation with silver; preparation of red wax; preparation of pepper (fol. 55 *verso*)
(68) preparation of amalgams; melting in the crucible (fol. 56 *recto*)
(69) operation with pomegranate peel; operations for whitening (fol. 56 *verso*)
(70) preparation of sharp vinegar; operation with sulphur (fol. 57 *recto*)
(71) operation on iron filings (fol. 57 *verso*)
(72) operation with mercury and arsenic; operation with sal ammoniac, vitriol, and borax (fol. 58 *recto*)
(73) operation with sal ammoniac, vitriol, sulphur, and egg white (fol. 58 *verso*)
(74) operations with tin and mercury (fol. 59 *recto*)
(75) purification of the spirit of silver with tin (fol. 59 *verso*)
(76) operation with iron and salt (fol. 60 *recto*)
(77) operation with iron filings and sal ammoniac (fol. 60 *verso*)
(78) operation with water of the eagle, the eagle, and arsenic; operation with eggshells, alkali salt, vitriol, salt of Andara and kitchen salt (fols. 61 *recto* & *verso*)

(79) operation with mercury, eagle, and iron filings; oil of the eagle; reddening of vitriol; making of the water of the servant (fol. 62 *recto*)
(80) mixing the servant with tutty; water of vitriol and scorpion (fol. 62 *verso*)
(81) operation with yellow poison; operation with egg; operation with a spirit, mercury, vitriol, sulphur, verdigris, and cinnabar (fol. 63 *recto*)
(82) purification of the spirit; operation with kitchen salt, alkali salt, and salt of borax; operation with red vitriol, the scorpion, kitchen salt, and verdigris (fol. 63 *verso*)
(83) operation with sublimed servant for making the Sun; operation with the servant, the eagle, and the scorpion (fol. 64 *recto*)
(84) operation with arsenic (fol. 64 *verso*)
(85) operation with the eagle; operations with eggs (fol. 65 *recto*)
(86) operation with iron filings; preparation of an amalgam (fol. 65 *verso*)
(87) operation with a spirit, sulphur, mercury, verdigris, cinnabar, and red vitriol; operation with iron filings (fol. 66 *recto*)
(88) operation with spirit, the scorpion, the servant, antimony, verdigris, and red vitriol (fol. 66 *verso*)
(89) operation with arsenic and borax (fols. 67 *recto* & *verso*)
(90) operation with iron filings; operation with arsenic, sulphur, and salt (fol. 68 *recto*)
(91) preparation of ashes from which a water is obtained; operation with mercury, the scorpion, the eagle, and alum (fol. 68 *verso*)
(92) operation with mercury, sal ammoniac, iron filings (fol. 69 *recto*)
(93) melting of the Moon; operation with lime, alkali salt, and the scorpion; clay of wisdom; the elixir (fol. 69 *verso*)
(94) calcination of Jupiter; operation with the Moon, alum, the servant, and tin; cleaning of arsenic in cow's milk (fol. 70 *recto*)
(95) operation with exhalations of tutty, mercury, and the scorpion; operation with lime, borax, the eagle, and salt; operation with verdigris and antimony (fol. 70 *verso*)
(96) operation on tin; operation on the servant (fol. 71 *recto*)
(97) operation with the eagle, borax, and salt of Andara; operation with iron filings (fol. 71 *verso*)
(98) operation with the slave, copper filings, the scorpion, bitter salt, and white vitriol; making of lime with eggshells (fol. 72 *recto*)
(99) operation with the scorpion, the eagle, verdigris and antimony; operation with white iron and the Moon (fol. 72 *verso*)
(100) operation with silver and iron; congealment of the servant (fol. 73 *recto*)
(101) operation with antimony, verdigris, and the eagle (fol. 73 *verso*)
(102) operation with bread salt, the scorpion, and vitriol; melting of lead (fol. 74 *recto*)
(103) 'killing' of the servant with sulphur; melting of white lead; operation with filings of Mars (fol. 74 *verso*)

(104) operation with tin and egg white; operation with calx and egg white; operation with the servant and water of the scorpion; operation with tartar (fol. 75 *recto*)
(105) operation with mercury and lead; operation with the scorpion and litharge; operation with vinegar of the root of celandine (fol. 75 *verso*)
(106) washing of lime with grease; *lacuna* (fol. 76 recto)
(107) *lacuna* (fols. 76 *verso*-77 *verso*)
(108) operation with the servant and filings of the Moon; operation with mercury, verdigris, and the scorpion (fol. 78 *recto*)
(109) operation with antimony, verdigris, and borax (fols. 78 *verso* & 79 *recto*)
(110) operations for mixing the Moon and the Sun (fol. 79 *verso*)
(111) operation with mercury; operation with yellow arsenic (fol. 80 *recto*)
(112) preparation of water of copper; melting of vitriol; operation with borax and iron filings (fol. 80 *verso*)
(113) operation with arsenic; operation with iron and copper filings (fol. 81 *recto*)
(114) operation with verdigris (fol. 81 *verso*)
(115) appeal to the secrecy of the operations described (fol. 82 *recto*)
(116) operation with ground sal ammoniac and ground verdigris (fol. 82 *verso*)
(117) operation with eggs and eggshells (fol. 83 *recto*)
(118) melting of the Sun (fol. 83 *verso*)
(119) operations with antimony, the scorpion, and verdigris (fol. 84 *recto*)
(120) operation with red arsenic, the scorpion, and the slave; grinding of antimony (fol. 84 *verso*)
(121) operation with cow's bile and honey; operation with arsenic (fol. 85 *recto*)
(122) operation with red lead and ground glass; operation with antimony, the scorpion, copper, and the eagle (fol. 85 *verso*)
(123) operations with white lead, silver, soap, and oil (fol. 86 *recto*)
(124) separation of copper; operation with tin, vitriol, and vinegar; increase of the weight of the scorpion and arsenic (fol. 86 *verso*)
(125) operation with the oil of the scorpion; operation with pure mercury; operation on the stone found in cow's dung (fol. 87 *recto*)
(126) operation with verdigris, silver filings, and white lead; operation with antimony and silver (fol. 87 *verso*)
(127) operation with arsenic; operation with cinnabar (fol. 88 *recto*)
(128) clay of wisdom, verdigris, and ashes; operation with cinnabar and pure verdigris (fol. 88 *verso*)
(129) operation with cinnabar and verdigris (fols. 89 *recto* & *verso*)
(130) [German indication of the number of folios in the manuscript, dated March 13, 1975] (fol. 90 *recto*)

Appendix 3. Comparative Table of Contents of the Different Versions of *On Alums and Salts*

The following table shows the relationship between the sections the Hebrew (H), the Arabic (A), and the Latin versions of the *OAS* (L^A: edition Arbuthnot 2002; L^R: edition Ruska 1935; L^S: edition Steele 1929). A dash (–) indicates the absence of the section in the corresponding version. No extant manuscript includes all the topics listed in the first column of the table: this list is simply to establish the relationship between the different versions of the *OAS*. As such, the first column represents the *OAS* in its most expanded version, by including sections that probably did not belong to the original treatise but became embedded in the Hebrew, the Arabic, or the Latin version during the history of the work's transmission.

Topic	H	A	L^A	L^R	L^S
On atrament	§ 1	–	§§ 1, 2	§§ 72, 73	§§ 1, 2
On alums	§ 2	§ 1 (partial)	§§ 7, 8	§§ 74, 75	§§ 3, 4
On salts	§ 3	§ 2	§§ 3–5	§ 76	§ 5
Preparation of sal ammoniac	§ 4	§ 3	§ 6	§ 77	§ 6
On alkali salt	§ 5	§ 4	§§ 9–11	§ 78	§ 7
Preparation of alkali salt	§ 6	§ 5	§ 12	§§ 78, 79	§ 8
Preparation of borax	§ 7	§ 6	§§ 13, 14	§ 80	§ 9
Preparation of an elixir from silver, mercury, salt, and eggshells	§ 8	–	–	–	–
On sal ammoniac	§ 9	§ 7	§§ 15, 16	§ 81	§ 9
Preparation of sal ammoniac (i)	§ 10	§ 8	§ 17	§ 82	§ 10
Preparation of sal ammoniac (ii)	§ 11	§ 9	§ 17	§ 83	§ 10
On Arsenic	§ 12	§ 10	§§ 18, 19	§ 1	§ 11
Sublimation of arsenic	§ 13	§ 11	§ 20	§ 2	§ 12
Improvement of arsenic (i)	§ 14	§ 12	§ 21	§ 3	§ 13
Improvement of arsenic (ii)	§ 15	§ 13	§ 22	§ 4	§ 14
Making silver with arsenic (i)	§ 16	§ 14	§§ 23, 24	§ 5	§ 15

Topic	H	A	L^A	L^R	L^S
Making silver with arsenic (ii)	§ 17	–	§ 25	§ 6	§ 16
On sulphur	§ 18	§ 15	§ 26	§ 7	§ 17
Preparation of sulphur	§ 19	§ 16	§ 27	§ 8	§ 18
Making gold with vitriol, iron, and copper	–	§ 17	–	–	–
Making gold with burnt copper	–	§ 18	–	–	–
Making silver with mercury, vinegar, and arsenic	–	§ 19	–	–	–
Melting glass	–	§ 20	–	–	–
Rectification of *qala'ī*]	–	§ 21	–	–	–
Making silver with talk, mercury, and borax	–	§ 22	–	–	–
Preparation of sugar	–	§ 23	–	–	–
Preparation of artificial musk	–	§ 24	–	–	–
On mercury	§ 20	–	§§ 28–31	§ 9	§ 19
Sublimation of mercury (i)	§ 21	–	§ 32	§ 10	§ 20
Sublimation of mercury (ii)	§ 22	–	§ 33	§ 11	§ 21
Congealment of mercury (i)	§ 23	–	§ 34	§ 12	§ 22
Congealment of mercury (ii)	§ 24	–	§§ 35, 36	§ 13	§ 23
Dissolution of mercury with sal ammoniac	–	–	§ 36	–	–
Congealment of mercury (iii)	§ 25	–	§ 37	§ 14	§ 24
Congealment of mercury (iv)	§ 26	–	§ 38	§ 15	§ 25
Nutrition with the bodies	§ 27	–	§§ 39, 40	§ 16	§ 26
On the bodies	§ 28	–	§§ 41–43	§§ 17, 18	§§ 27–28
Ceration of the bodies	§ 29	–	§§ 44, 45	§§ 19, 20	§§ 29–30

Topic	H	A	L^A	L^R	L^S
Congealment of the bodies	§ 30	–	§ 46	§ 21	§ 31
On gold	§ 31	–	§§ 47–49	§§ 22, 23	§§ 32, 33
Preparation of arsenic	§ 32	–	§ 50	§ 24	§ 34
Preparation of gold (i)	§ 33	–	§ 51	§ 25	§ 35
Preparation of gold (ii)	§ 34	–	§ 52	§ 26	§ 36
Preparation of gold (iii)	§ 35	–	§ 53	§ 27	§ 37
Preparation of gold (iv)	§ 36	–	§ 54	§ 28	§ 38
Preparation of gold (v)]	§ 37	–	§ 55	§ 29	§ 39
Preparation of gold (vi)	§ 38	§ 25 (partial)	§ 56	§ 30	§ 40
On silver	§ 39	§ 26	§§ 57–60	§§ 31, 32	§§ 41, 42
Preparation of silver (i)	§ 40	§ 27	§ 61	§ 33	§ 43
Preparation of silver (ii)	§ 41	§ 28	§ 62	§ 34	§ 44
Preparation of silver (iii)	§ 42	§ 29	§ 63	§ 35	§ 45
Preparation of silver (iv)	§ 43	§ 30	§ 64	§ 36	§ 46
Preparation of silver (v)	§ 44	§ 31	§ 65	§ 37	§ 47
Preparation of silver (vi)	§ 45	§ 32	§ 66	§ 38	§ 48
On iron	§ 46	§ 40	§ 67	§ 39	§ 49
Preparation of iron (i)	§ 47	§ 41	§ 68	§ 40	§ 50
Preparation of iron (ii)	§ 48	§ 42	§§ 69–71	§§ 41–42	§§ 51–53
Preparation of iron (iii)	§ 49	§ 43	§ 73	–	§ 53
Preparation of iron (iv)	–	§ 44	§ 74	§ 44	§ 54
Preparation of yellow sulphur	§ 50	§ 45	§ 75	–	–
Preparation of iron (v)	–	§ 46	–	–	–
On copper	–	§ 47	§ 80	§ 48	§ 58
Preparation of copper (i)	§ 51 (partial)	§ 48	§ 81	§ 49	§ 59

Topic	H	A	L^A	L^R	L^S
Preparation of copper (ii)	§ 52	§ 49	§ 82	§ 50	§ 60
Cleaning of iron and copper	§ 53	–	§ 86	§ 53	§ 63
Making silver with copper and mercury	§ 54	§ 50	§ 87	§ 54	§ 64
On *qala'ī*/tin	§ 55	§ 33	§ 88	§ 55	§ 65
Making silver from *qala'ī*/tin (i)	§ 56	§ 34	§ 89	§ 56	§ 66
Making silver from *qala'ī*/tin (ii)	–	§ 35	§ 90	–	–
Making silver from *qala'ī*/tin (iii)	§ 57	§ 36	§§ 90, 91	§ 57	§ 67
On lead	§ 58	§ 37	§§ 92–94	§ 58	§ 68
Preparation of lead (i)	§ 59	§ 38	§ 95	–	–
Preparation of lead (ii)	§ 60	§ 39	§ 96	§ 60	–
Preparation of lead (iii)	§ 61	–	§ 97	§ 61	§ 69
On the white egg	§ 62	–	§ 98	–	–
On glass	§ 63	§ 51	§ 100	§ 64	–
Preparation of artificial ruby (i)	§ 64	§ 52	§ 101	§ 65	–
Preparation of an artificial yellow gemstone	–	§ 53	§ 102	§ 66	–
Preparation of an artificial azure gemstone	§ 65	§ 54	§ 103	§ 67	–
Preparation of an artificial green gemstone	§ 66	§ 55	§ 104	§ 68	–
On the ninth and its preparation (i)	§ 67	–	§ 106	–	–
Preparation of the ninth (ii)	§ 68	–	§ 107	–	–
On marcasite	§ 69	–	§ 108	–	–
Preparation of marcasite	§ 70	–	§ 108	–	–

Appendix 4. Latin Manuscripts of *On Alums and Salts*

In the current state of our knowledge, seven Latin manuscripts that preserve the *OAS* are known. These manuscripts represent at least two different translations made from the Arabic text by different translators at different times. While we know that the translator of one of these two versions was Gerard of Cremona, the authorship and chronological setting of the second is unknown. A critical edition and English translation of the Latin *OAS* were presented in Catherine Arbuthnott's doctoral thesis and awaits publication.[516]

The following manuscripts constitute the extant evidence of the Latin translations of the *OAS*:

1. MS 4 Qq A 10, Palermo, Biblioteca Comunale (aka 'Codex Speciale'), beginning of the XIV century, parchment, fols. 233r-244v. The manuscript is written in minute but clear handwriting and preserves a collection of some of the most famous medieval alchemical works.[517] Carini proposed dating the manuscript between the thirteenth and fourteenth centuries on the basis of the fact that in some of the treatises Thomas Aquinas is mentioned as a friar or as consecrated, while in others he is defined as a saint. Carini moreover hypothesizes that the compiler of the codex might be a certain Friar Domenico of the San Procolo monastery in Bologna, who is considered by Carini the owner of most of the manuscripts anthologised in the Palermitan codex.[518] The *OAS* has the title *Sermo de aluminibus et salibus, quae in haec arte necessaria existunt*. A marginal note offers a peculiar alternative title: *Alii intitulant hunc ita: Incipit liber Ypocratis et Galieni*. The incipit of the *OAS* in this manuscript reads: *Scias quod atramenti sunt genera multa* and the work is concluded by the explicit *aqua ergo ecsir indiget illa aqua intellige. Explicit liber. Deo gratias*. Carini provides an inventory of the contents of the Palermo manuscript, where the *OAS* is listed as number 34, but does not provide any indication of foliation.[519] The codex used to be part of the collection of books and manuscripts of the family Speciale from Nicosia, one of the most illustrious and ancient Sicilian families. Famous members of this family include Niccolò Speciale, viceroy of the kingdom between 1423 and 1432; another Niccolò, who authored eight books on Sicilian history; and Pietro Speciale, magistrate of Palermo until 1469. Between 27 and 30 May 1860, during the insurrection that led to the end of the Bourbon control over the city and eventually to the unification of Sicily with the newly-born Italian kingdom, the Palermitan palace of the brothers Gaetano and Pietro Speciale was pillaged by solders and civilians, and a large part of their manuscript

[516] Arbuthnott, *Pseudo-Razi*. My notes on the Latin versions of the *OAS* are based on the material presented in Arbuthnott's work and on exchanges on the topic with Professor Charles Burnett.

[517] See Carini, *Sulle scienze occulte*, 5–8. A detailed description of the manuscript is in Unione Accademica Nazionale, *Catalogo dei manoscritti filosofici delle biblioteche italiane*, 11 vols. (Florence: Olschki, 1993), VII, 97–105.

[518] Carini, *Sulle scienze occulte*, 8. Carini does not provide enough evidence to consider his idea more than just an interesting opinion.

[519] Carini, *Sulle scienze occulte*, XIV.

collection was lost or destroyed. Luckily, the 'Speciale' codex survived this fate; in 1872 Carini wrote that the codex was on the market and hoped it would be bought by Palermo's Municipal Library. This happened in 1873. In his description of this witness of the *OAS*, Carini provides a list of the section headings that shows that the structure of this version of the *OAS* coincides with the order of topics found in the Arabic Sprenger MS 1908 and the Hebrew MS Orient. 514; the same structure was also preserved in Steele's 1929 edition.

2. MS Lat. 6514, Paris, Bibliothèque Nationale, XIV-XV century, parchment, fols. 125r-128v.The manuscript appears to preserve the Latin translation by Gerard of Cremona. The author of the *OAS* is here called 'Bubacher Magumet fil. Ceceri Arrasi,' and this indication was crucial for the circulation of the work and its fame as an original work by al-Rāzī. Berthelot's study of the manuscript is marred by some errors. For example, he maintained that the explicit of the *OAS* reads "*explicit liber fratris Rogeri Bachonis.*"[520] This explicit actually belongs to Bacon's treatise *Breve breviarium*, but Berthelot did not notice that the last four folios of the *OAS*, where the genuine explicit of the treatise and the beginning of Bacon's *Breviarium* would have been found, had fallen out from the manuscript.[521] Berthelot's confusion might have been encouraged by the similarity of topics between the *OAS* and the *Breve breviarium* which was strongly influenced by the pseudo-Avicennian *De Anima*.[522] Berthelot also argues that the quotations of the *OAS* that appear in Vincent of Beauvais' works cannot be identified (apart from a single case) in the Parisian manuscript; this assertion was disproved by Steele who argued that those sections are indeed present and that at least six of them are found in this manuscript, but escaped Berthelot's eye.
3. MS Digby 119, Oxford, Bodleian Libraries, XIV century, parchment, fols. 167v-175v. A marginal note, possibly in the hand of John Dee, provides the title *Liber rationum super corpora, spiritus, sales et atramenta*.[523]
4. MS Corpus Christi Coll. 125, Oxford, Corpus Christi College, XIV century, parchment, fols. 110r-117r. The first folio of the manuscript preserves a note of possession dated 1592: *Liber Thomas Sprot. de librari. Sancti Agustini Canonorum*. The *OAS* is titled *Liber salium et alluminibus alchemiae* and ends with the explicit: *et sic suscipit duriciam er albedinem donec veniati in quantitate lune et sic erit optimum in operacione transmutacionis et ad faciendum magnum cumulum azimum id est salem*.[524]

[520] Berthelot, *La Chimie*, I, 317–19.
[521] See Steele, "Practical Chemistry," 10–11.
[522] For the critical edition, French translation and a thorough analysis of the *De Anima*, see Sébastien Moureau, *Le* De Anima *alchimique du pseudo-Avicenna*, 2 vols. (Florence: Sismel-Edizioni del Galluzzo, 2016).
[523] See above, § 3.1.
[524] Singer, *Catalogue*, 107–8.

5. MS El. q. 20, Thüringer Universitäts- und Landesbibliothek Jena, XIV century, parchment, fols 29r-41v.[525] The incipit of the *OAS* reads: "Incipit Liber Lumen Luminum. Atramentorum genera sicut [sic?] multa et eius minere sunt inuente."
6. MS 474 (830), Bologna, Biblioteca Universitaria, XV century, parchment, fols. 35r-48v. The manuscript, which is part of the collection of Count Caprara, is described in detail in the catalogue of the Latin codices of the library published by Lodovico Frati.[526] The *OAS* opens with the incipit: *Sermo de aluminibus et salibus et atramentis et alii quae in hac arte necessari existunt ad complementum eleizir.*
7. MS Arundel 164, London, British Library, XV century, paper, fols. 167v-175v. The text of the *OAS* is here surrounded by works attributed to the Latin Geber; it is titled *Tractatus de aluminibus et salibus et primo de atramento* and ends with the explicit: *scias ergo statione eius cum aqua regi rubei aliquotiens est enim ultimum ergo occulta ipsum et age cum eo et prosperaberis. Explicit secretum philosophiae secretissimum.* Steele used this manuscript as a counterpart to the Parisian manuscript 6514 for preparing his edition.

Two other manuscripts were originally believed to preserve Latin versions of the *OAS*, given that their titles and incipits closely resemble those of the Latin *OAS*; close analysis of the texts they preserve identified them as a different alchemical work:

A. MS 1400 (III), Cambridge, Trinity College, XV-XVI century, fols. 75v-81r. The treatise *Liber Saturni et aluminum* (whose title could easily have been the result of a copyist's error for *Liber salium et aluminum*) opens with an incipit closely resembling that of the *OAS*: *Salium autem sunt genera multa. Sal armoniacus, sal communis et marinus. Et hic duo sunt genera vel species. Nam grossior et nigrior prevalent, Sal gemme qui est coloris cristallini, Sal nitri. Sal vitri. Sal petre. Sal alkali. Sal calcis. Sal cineris. Sal urine. Urin. Sal enim omnis participat in caliditate.* The explicit reads: *uterusque cum alio sine aliquo addito et interdum expedit exponere salem et ad diversitatem operis intenti diversatur sal. Explicit.* [527] Ruska analysed this manuscript for his edition of the Latin *OAS* and discarded it as a work different from the *OAS*.

B. MS No. Hu 1051, San Marino (CA), Henry E. Huntington Library, XV–XVI century, fols. 110v-23v. The codex is an alchemical multiple-text manuscript in both Latin and English by at least thirteen different copyists who collected material attributed to major alchemical (and not strictly alchemical) authorities from

[525] The manuscript is available in digital version at the following address: https://collections.thulb.uni-jena.de/receive/HisBest_cbu_00015896 (accessed 14 October 2022). I am grateful to Sébastien Moureau who, having discovered this further version of the Latin *OAS*, shared its details with me while I was finalizing this section of the book.

[526] Lodovico Frati, *Indice dei codici latini conservati nella R. Biblioteca Universitaria di Bologna* (Florence: Successori Seeber, 1909; reprint of the article first published in *Studi italiani di filologia classica* 16 [1908]: 103–482), 1–171. For a presentation of the Caprara collection, see Didier Kahn, "Le fonds Caprara de manuscrits alchimique de la Bibliothèque universitaire de Bologne," *Scriptorium* 48 (1994): 62–110.

[527] Singer, *Catalogue*, 107–8.

Aristotle to the fifteenth century.[528] The treatise, that may be confused with another translation of the *OAS*, opens with *Incipit de salibus et eorum preparacione. Salium autem multa sunt genera, scilicet sal armoniacum, sal gemme, sal communis, sal marinus, et huius [nigrior] prevalent sal nitri, sal vitri, sal petre, sal alkali, sal calcis, sal cineris et sal urine*. This incipit appears to reproduce the opening of the treatise on salts preserved in the Trinity College manuscript (A) with some significant variants. The text does not represent another translation of the *OAS*, but rather an autonomous treatise upon which the Latin *OAS* may have exerted an influence.

The preliminary analysis of the extant witnesses of the Latin *OAS* led Arbuthnott to classify them in two distinct families: A, represented by MS Digby 119 and the text used in the John of Garland printed *Compendium alchimiae*, and family B, represented by the remaining five manuscript witnesses. The complex relationships between the two families and among the manuscripts that constitute them did not allow Arbuthnott to draw a solid *stemma codicum* and therefore her critical edition of the Latin *OAS* remains at times conjectural.

[528] For a detailed description of this manuscript, see William J. Wilson, "Catalogue of Latin and Vernacular Alchemical Manuscripts in the United States and Canada," *Osiris* 6 (1939): 419–61.

Appendix 5. Lexicon of Alchemical Terminology in the Arabic and Hebrew *On Alums and Salts*

A. Arabic-English

Arabic	English	Paragraph
آنك	Tin	§§ 21, 26, 39
أتون	Furnace	§§ 28, 29
أثال	Aludel	§ 9
أرض	Earth	§§ 2, 26, 37
أسرب	Lead	§§ 14, 26, 28, 33, 36, 37, 38, 39, 40, 41, 42, 43, 44, 46, 48
أسرنج	Red lead	§ 37
إسفيداج	White lead	§§ 37, 43, 44
إكسير	Elixir	§§ 7, 26, 35, 37, 47
إناء فخار	Terracotta vessel	§ 34
برادة	Filings	§§ 6, 17, 19, 29, 30, 37, 42, 46, 50
بلور	Crystal	§ 5
بودقة	Crucible	§§ 6, 52
بورق الحكماء	Borax of the Sages	§ 43
بوط	Crucible	§§ 27, 34, 41, 42, 43, 44, 48
بوط بر بوط	Double crucible	§ 42
بول	Urine	§§ 16, 17, 31, 34, 44, 48, 50
تجسد	Corporality	§ 2
تدبير	Operation	§§ 2, 7, 10, 11, 14, 15, 16, 22, 26, 27, 34, 38, 40, 41, 47
تنكار	Borax	§§ 6, 22, 43
توتيا	Tutty	§ 18
تصعيد	Sublimation	§§ 7, 10, 11
ثفل	Sediment	§§ 1, 5
جسد	Body (metal)	§§ 2, 4, 6, 7, 10, 13, 14, 26, 32, 35, 37, 40, 41, 46, 51, 55
جير	Lime	§ 21
حجر	Stone	§§ 2, 4, 6, 7, 10, 22, 35
حديد	Iron	§§ 7, 19, 34, 35, 36, 37, 39, 40, 41, 53
حرارة	Hotness	§§ 4, 10, 33, 37, 47
حل	Dissolution	§§ 1, 3, 7, 12, 13, 20, 25, 38, 46, 50
حيوان	Animal	§ 2
حيوانية	Animality	§§ 4, 7

Arabic	English	Paragraph
خبث	Dross	§ 42
خل	Vinegar	§§ 6, 10, 13, 19, 34, 36, 38, 39, 46
دم	Blood	§§ 24, 25
دم اخوين	Dragon's blood	§ 18
دهن	Oil	§§ 4, 7, 10, 14, 42
دهنج	Malachite	§ 55
دهانية/ دهنية	Unctuosity	§§ 4, 10, 15
دواء	Medicine	§§ 22, 34, 48, 49
ذهب	Gold	§ 26
ذوب	Fusion	§§ 10, 33, 37, 42, 43, 47
رائحة	Exhalation	§ 37
راسخت	*Rāsḫat*	§§ 15, 18, 49
رصاص	Lead	§§ 8, 22
رطوبة	Humidity	§§ 4, 7, 10, 33, 37
رماد	Ash	§§ 2, 30
روح	Spirit	§§ 4, 6, 7, 10, 26, 55
روحانية	Spirituality	§ 2
زاج	Vitriol	§§ 17, 25, 46, 47, 48, 49, 50, 53
زاج قبرصي	Cypriot vitriol	§ 18
زجاج	Glass	§§ 4, 20, 42, 43, 51, 52, 53, 54, 55
زجاجة	Flask	§§ 14, 17, 46
زرنيخ	Arsenic	§ 35
زعفران	Saffron	§§ 18, 46, 53
زنجار	Verdigris	§§ 49, 55
زنجفر	Cinnabar	§§ 15, 18, 46, 48
زهرة	Venus (copper)	§§ 10, 14, 15, 17, 22, 26, 33, 34, 35, 37, 47, 48, 50
زيبق	Mercury/quicksilver	§§ 10, 22
شاذنة	Hematite	§ 52
شب	Alum	§§ 31, 32, 40, 42, 47, 54
شب العصفر	Safflower alum	§ 5
شب يماني	Yemeni alum	§§ 16, 21, 34, 50
شحم	Fat	§§ 6, 19, 32
شعر	Hair	§ 18
شقف	Pot sherd	§§ 30, 45

Arabic	English	Paragraph
شمس	Sun (gold)	§§ 2, 8, 26, 33, 36, 37, 40, 46, 47, 48, 49
صابون الحكماء	Soap of the sages	§ 2
صاروج	Quicklime	§ 11
صبغ	Tincture/dye	§§ 10, 26, 33, 37, 40, 47
صفحة	Sheet/plate	§§ 13, 31, 32, 41, 43, 44, 45, 46, 48
صلاية	Mortar	§§ 34, 38, 47
طبع	Nature	§§ 2, 4, 6, 7, 10, 15, 26, 33, 40, 47, 51
طلق	Talc	§ 22
طين الحكمة/ طين المحكم	Clay of the Sages	§§ 31, 32, 34, 42, 46
طين الشعر	Clay of the hair	§ 11
عبد	Servant (Mercury/quicksilver)	§§ 4, 7, 10, 15, 19, 20, 22, 25, 26, 33, 37, 38, 41, 50
عظم	Bones	§§ 2, 44
عقاب	Eagle (sal ammoniac)	§§ 7, 9, 20, 26, 31, 32, 36, 38, 40, 42, 46, 48, 49, 50
عقاب مصري	Egyptian eagle (Egyptian sal ammoniac)	§ 34
عقرب	Scorpion (sulphur)	§§ 4, 10, 15, 28, 29, 30, 31, 33, 37, 39, 45, 46, 48, 49
علم	Sign (arsenic)	§§ 4, 10, 11, 13, 14, 15, 19, 26, 28, 31, 32, 39, 40, 42, 43, 48
غبار	Powder	§§ 6, 31, 32, 34, 41
فرن	Oven	§§ 17, 41, 42, 48
فرن الخبز	Bread oven	§§ 3, 12, 14, 32, 34, 46, 48, 49
فضة	Silver	§§ 29, 31, 33, 37
فهر	Pestle	§ 34
فيروزج	Turquoise	§ 54
قدح	Bowl	§ 22
قدر/قدرة	Cooking pot	§§ 3, 5, 7, 8, 32
قدر مزجج / قديرة مزججة	Glazed pot	§§ 1, 11, 12, 39, 46, 49
قطران	Tar	§§ 21, 36
قلعي	*Qalaʿī*	§§ 17, 20, 21, 26, 27, 28, 32, 33, 35, 36, 37, 40, 43, 46
قمر	Moon (silver)	§§ 2, 8, 14, 25, 26, 28, 32, 33, 34, 37, 40, 43, 46, 47, 48, 50
كافور	Camphor	§ 11

Arabic	English	Paragraph
كبريت	Sulphur	§ 26
كحل	Kohl	§ 22
كركم	Turmeric	§ 18
كلس	Calx	§§ 2, 29, 34, 39
كوز	Jar	§ 48
كير	Bellows	§§ 41, 42
كيمياء	Elixir/alchemy	§ 36
لازورد معدني	Mineral lapis lazuli	§ 54
ماء الجبن	Whey	§§ 6, 16
مرتك	Litharge	§ 37, 43
مرقشيثا	Marcasite	§§ 22, 40
مريخ	Mars (iron)	§§ 5, 8, 14, 17, 26, 33, 37, 40, 41, 42, 43, 44, 45, 46, 47, 51
مزاج	Mixture/composition	§§ 4, 33
مسك	Musk	§ 24
معدن معدنية	Mineral/ore	§§ 4, 10, 33, 37
مغرفة حديد	Iron spoon	§§ 34, 35, 36, 38, 39, 41
مغنيسيا	Magnesia	§§ 37, 40, 54
مقرعة	Stick	§§ 19, 20
ملح	Salt	§§ 2, 4, 5, 6, 7, 9, 11, 12, 25, 28, 29, 30, 31, 34, 36, 38, 39, 48, 50
ملح الأندران	Salt of Andara	§ 2
ملح الحيوان	Animal salt	§ 4
ملح الخبز	Bread Salt	§§ 2, 6
ملح الطعام	Kitchen salt	§§ 2, 11, 14
ملح العجين	Dough salt	§§ 4, 7, 13
ملح القلي	Alkali salt	§§ 4, 12, 42, 43, 44, 53
ملح مر	Bitter salt	§ 2
ملح هندي	Indian salt	§ 2
مهراس	Mortar	§ 49
نبات	Plant	§§ 2, 4
نباتي	Vegetal	§ 4

Arabic	English	Paragraph
نحاس	Copper	§§ 5, 8, 26, 48, 49
نشادر	Sal ammoniac	§§ 6, 7, 20, 23
نطرون	Natron	§ 20
نفس	Soul	§§ 10, 26
نورة	Lime	§ 10
هاون	Mortar	§§ 14, 48
يبس	Dryness	§§ 2, 4, 7, 37, 47, 51
ياقوت	Gemstone	§ 53

B. Hebrew-English

Hebrew	English	Paragraph
אבן	Stone	§§ 2, 3, 5, 8, 12, 20, 31, 63, 67
אבן אזוורד	Lapis lazuli	§ 65
אבק	Dust	§§ 7, 13, 36, 47
אוגיינטו/אוניינטו	Oily substance/ ointment	§§ 2, 5, 9, 20, 28
אופירציאוני	Operation	§ 6
אטרמנטו	Atrament/vitriol	§§ 1, 20, 24, 33, 50, 52, 54, 64
איקציר/אלקציר/ אדלסיר/אקציר	Elixir	§§ 8, 9, 20, 31, 39, 48, 53, 58, 70
אלביט/בוט	Crucible	§§ 37, 40, 48, 49, 51, 53, 56, 64, 70
אלוטילו	Aludel	§§ 11, 21, 22
אלומי	Alum	§§ 2, 17, 22, 27, 44, 45, 53, 65, 70
אלומי דאלסור	Alum *sory*	§ 6
אלומי יימיני	Yemeni alum	§§ 19, 54, 56
אלוק/אלונקי	Tin	§§ 17, 55, 60, 61
אלמליאה/אנימלי	Animal	§§ 5, 9, 20
אמיטיט	Amethyst	§ 64
אנימליטטי	Animality	§ 5
אנפולא	Vial	§ 54
אנפולה אטורטא	Twisted vial	§ 45
אסדנג	White lead	§ 58
אסרוב	Lead	§§ 25, 58, 60, 61
אפר	Ash	§§ 3, 9, 16, 30, 43, 45, 54, 57, 60, 67, 68
בדיל	Tin/*qala'T*	§§ 3, 10, 11, 16, 17, 20, 27, 28, 33, 39, 44, 45, 53, 54, 55, 56, 57, 58, 68, 70
בורק	Borax	§ 70
בייטבלי/בייטיבלי	Vegetal	§ 5
ברזל	Iron	§§ 9, 10, 16, 25, 27, 28, 43, 45, 46, 47, 48, 49, 50, 53, 56, 57, 58, 59, 60, 63, 64, 69
גחל	Coal	§§ 16, 17, 33, 43, 44, 48, 50
גוף	Body (metal)	§§ 1, 2, 3, 5, 7, 8, 9, 12, 15, 20, 27, 28, 29, 31, 32, 39, 45, 46, 47, 58, 63, 66
גפרית	Sulphur	§§ 1, 2, 3, 12, 18, 20, 23, 24, 28, 31, 33, 35, 36, 39, 41, 42, 43, 50, 52, 58, 60
דיתריירו	Litharge	§ 49
דם	Blood	§ 5
זאג	Vitriol	§§ 1, 9, 21, 36, 54, 70
זבל	Manure	§§ 27, 29, 44
זהב	Gold	§§ 3, 10, 20, 28, 31, 33, 34, 35, 36, 37, 38, 39, 46, 51, 52, 54, 55, 58
זנייר/ זינגייר	Verdigris	§§ 52, 66
זכוכית	Glass	§§ 5, 10, 23, 29, 30, 33, 38, 48, 54, 62, 63, 64, 65, 70
זרניך	Arsenic	§§ 2, 11, 12, 13, 15, 16, 17, 18, 20, 27, 31, 32, 33, 35, 39, 41, 44, 45, 46, 48, 49, 59, 60, 68

Hebrew	English	Paragraph
חומץ	Vinegar	§§ 7, 12, 15, 16, 20, 27, 32, 38, 53, 54, 57, 59, 60, 61, 62, 70
חלב הבתולה	Milk of the virgin	§§ 20, 57
חמוץ	Ferment	§ 2
חמימות	Hotness	§§ 28, 29, 31, 50, 58
טוטיאה	Tutty	§ 70
טיט	Clay	§§ 24, 29, 45
טיט החכמה	Clay of the Sages	§§ 12, 13, 17, 25, 26, 35, 44, 48, 50, 56
טלק	Talc	§ 5
טנכר	Borax	§§ 7, 48, 53
טס	Sheet	§§ 34, 36, 41, 44, 45, 47, 48, 49, 50, 51, 53, 54, 58, 62
יבשות	Dryness	§§ 31, 44
יימי ירוקות	Green gemstones	§ 66
יימי רוביאי	Red gemstones	§ 65
יקווינטו	Hyacinth	§ 64
כוכב	Star (Mercury/ quicksilver)	§§ 3, 8, 9, 18, 20, 21, 22, 23, 24, 25, 26, 27, 31, 33, 36, 38, 39, 44, 45, 47, 48, 49, 54, 55, 56, 58, 59, 63, 67, 70
כור	Furnace/crucible	§§ 43, 54
כור הצורפים	Goldsmith's furnace	§ 33
כלי	Vessel	§§ 2, 10, 13, 14, 23, 28, 29, 30, 35, 37, 38, 47, 48, 53, 56, 58, 62
כסף	Silver	§§ 3, 10, 13, 16, 17, 18, 27, 28, 38, 39, 40, 41, 42, 43, 44, 45, 46, 48, 49, 51, 52, 53, 54, 54, 55, 56, 58, 67, 69, 70
כסף הפילוסופים	Philosophers' silver	§ 3
כסף חי	Quicksilver/mercury	§§ 12, 17, 20, 28, 67
לבנה	Moon (silver)	§§ 8, 9, 12, 21, 28, 29, 30, 31, 39
לחות	Humidity	§§ 8, 31, 39
לימטורא/לימטורי/ לימא	Filings	§§ 7, 27, 35, 37, 39, 54, 62
למנא	Foil/sheet	§§ 7, 13
מגניסיאה	Magnesia	§§ 3, 46, 65
מזוכך	Cooking pot	§§ 6, 7, 10, 14, 16, 17, 29, 50, 56, 67, 68
מכתש	Mortar	§ 52, 53
מלגמא	Amalgam	§ 8
מלח	Salt	§§ 3, 5, 7, 8, 9, 10, 11, 13, 14, 16, 20, 21, 22, 27, 28, 32, 33, 35, 36, 38, 41, 42, 43, 44, 48, 53, 54, 55, 56, 57, 59, 60, 61, 62, 66, 68
מלח אלקלי	Alkali salt	§§ 5, 9, 14, 24, 48, 49, 64, 68, 70
מלח ארמוניאקו	Sal ammoniac	§§ 3, 9
מלח בורק	Borax salt	§ 49
מלח האפר	Salt of ashes	§ 9
מלח הדם	Salt of blood	§ 3
מלח החזק	Strong salt	§ 3

Hebrew	English	Paragraph
מלח הלחם	Bread salt	§§ 3, 7, 9, 17, 57
מלח הנדי	Indian salt	§ 3
מלח הסיד	Salt of calx	§§ 3, 9
מלח מר	Bitter salt	§ 3
מלח משי	Dough salt	§§ 5, 15
מנירים/ מינירלי/ מינירליאה	Mineral	§§ 12, 20, 28
מנירה/ מנירים/ מנירא	Mine	§§ 1, 5, 11, 28, 55, 58
מעדן	Mineral/ore	§ 12
מפוחים/מפוח	Bellows	§§ 33, 47, 48, 56
מרקשיתא	Marcasite	§§ 46, 69
מרתך/ מרתק	Litharge	§§ 34, 37, 58
נחושת/ נחשת	Copper	§§ 9, 10, 16, 17, 18, 27, 28, 35, 39, 44, 48, 52, 53, 54, 55, 56, 58, 62, 69, 70
נטרו/נטרון	Natron	§§ 48, 49, 53, 70
נפש	Soul	§§ 11, 12, 39
נשדרא	Sal ammoniac	§§ 7, 9, 11, 20, 22, 24, 27, 33, 36, 38, 39, 44, 45, 46, 49, 52, 54, 56, 57, 59, 67, 70
סולימטו	Sublimed	§§ 12, 20, 24
סיירו	Whey	§§ 17, 19
סיד	Calx	§§ 3, 7, 8, 12, 16, 27, 28, 31, 35, 38, 56, 58, 60
סיר	Cooking pot	§ 9
סל אלקלי	Alkali salt	§§ 5, 48, 49
סל יימא	Rock salt	§ 3
ספוגייא	Sponge	§ 13
ספון	Soap	§ 67
ספון החכמים	Soap of the Sages	§ 3
עופרת/עפרת	Lead	§§ 17, 23, 25, 26, 33, 36, 39, 39, 46, 47, 48, 49, 53, 58, 59, 62, 69
עוקב	Eagle (sal ammoniac)	§ 1
עלאה	Sublimation	§§ 12, 13, 21, 22
עפר	Earth/dirt	§§ 1, 3, 7, 31, 45
עצמות	Bones	§§ 3, 5
פטיש	Hammer	§§ 48, 53, 70
צבע/ צביעה	Tincture/dye	§§ 1, 2, 24, 39, 46, 54, 55, 58, 63
צנברו	Cinnabar	§§ 18, 24, 30

Hebrew	English	Paragraph
ציפרו	Copper	§ 17
צירוסי	Ceruse	§§ 49, 58, 62
קדרה	Pot	§§ 4, 23, 24, 25, 30, 33, 45
קדרה מזוככת	Glazed pot	§§ 2, 13, 50, 52, 60, 67
קונדימנטו	Operation/ preparation	§§ 1, 28, 49, 56, 67, 69
קלס	Calx	§§ 28, 42, 67
קלעי	*Qalaʿī*	§§ 39, 40, 41, 48, 49, 55, 58, 60, 67, 69
קנפורא / כנפורא	Camphor	§§ 13, 30
קערה	Bowl	§ 67
קערה מזוככת	Glazed bowl	§§ 58, 67
קרוקו של ברזל	*Crocus* of iron	§§ 24, 64
קריסתלו/ קריסטלו	Crystal	§§ 6, 7
קרירות	Coldness	§§ 31, 58
רובינו	Ruby	§ 64
רוח	Spirit (volatile substance)	§§ 7, 8, 9, 10, 20, 39
רוחות	Spirituality	§§ 3, 9, 11, 24, 27, 57, 66
ריח	Exhalation	§§ 20, 24, 25, 26, 58, 70
רפואה	Medicine	§§ 9, 20, 30, 51, 52, 56, 64
שאור	Ferment	§§ 20, 31
שדנא	amethyst	§§ 3, 64
שומן	Greasiness	§§ 12, 18, 19
שילפו / שולפו	Sulphur	§§ 50, 52
שמן	Oil	§§ 23, 37, 44, 47, 48, 50, 70
שמן זית	Olive Oil	§ 23
שמן שקדים מרים	Bitter almond oil	§ 12
שמרים	Sediment	§§ 2, 6, 50, 62
שמש	Sun (gold)	§§ 20, 27, 55
שעוה	Wax	§§ 8, 25, 29, 44, 45, 59, 68
שער	Hair	§§ 5, 12, 13
שתן	Urine	§§ 2, 5, 19, 20, 44, 50, 51, 54, 56
תנור	Oven	§§ 1, 4, 14, 16, 21, 41, 47, 48, 59
תנור הלחם	Bread oven	§§ 21, 45, 50, 52, 54, 56
תנור של יוצרים	Potters' oven	§ 33
תסעיד	Sublimation	§ 13
תרווד	Ladle	§§ 2, 56, 57, 59, 60, 62, 70

C. English-Arabic-Hebrew[529]

The following lexicon provides correspondences between the English translation of the *OAS* and the Arabic and Hebrew versions of the text. It is intended as a tool to help readers navigate my translations, and to indicate the range of terminology employed in the different versions of the work. It should however be noted that establishing a precise, one-to-one correspondence between Arabic and Hebrew terms is a risky endeavour at the best of times, and often impossible. Since these two manuscripts are not derived from a common exemplar, and nor is one a translation of the other, the correspondences between terms should be treated as approximate. Rather, the terminology reflects the whole history of a rich manuscript tradition that is, in large part, unknown to us.

English	Arabic	Hebrew
Alkali salt	ملح القلي §§ 4, 12, 42, 43, 44, 53	מלח אלקלי §§ 5, 9, 14, 24, 48, 49, 64, 68, 70 סל אלקלי §§ 5, 48, 49
Aludel	أثال § 9	אלוטילו §§ 11, 21, 22
Alum	شب §§ 31, 32, 40, 42, 47, 54	אלומי §§ 2, 17, 22, 27, 44, 45, 53, 65, 70
Alum sory	—	אלומי דאלסור § 6
Amalgam	—	מלגמא § 8
Amethyst	شاذنة § 52	אמיטיט § 64 שדנא §§ 3, 64
Animal	حيوان § 2	אלמליאה/אנימלי §§ 5, 9, 20
Animal salt	ملح الحيوان § 4	
Animality	حيوانية §§ 4, 7	אנימליטטי § 5
Arsenic	زرنيخ § 35	זרניך §§ 2, 11, 12, 13, 15, 16, 17, 18, 20, 27, 31, 32, 33, 35, 39, 41, 44, 45, 46, 48, 49, 59, 60, 68

[529] The numbers below each Arabic and Hebrew term indicate the sections of my edition and translation of the Arabic and Hebrew *OAS* in which they are found.

English	Arabic	Hebrew
Ash	رماد §§ 2, 30	אפר §§ 3, 9, 16, 30, 43, 45, 54, 57, 60, 67, 68
Atrament (see also Vitriol)	—	אטרמנטו §§ 1, 20, 24, 33, 50, 52, 54, 64
Bellows	كير §§ 41, 42	מפוחים/מפוח §§ 33, 47, 48, 56
Bitter almond oil	—	שמן שקדים מרים § 12
Bitter salt	ملح مر § 2	מלח מר § 3
Blood	دم §§ 24, 25	דם § 5
Body (metal)	جسد §§ 2, 4, 6, 7, 10, 13, 14, 26, 32, 35, 37, 40, 41, 46, 51, 55	גוף §§ 1, 2, 3, 5, 7, 8, 9, 12, 15, 20, 27, 28, 29, 31, 32, 39, 45, 46, 47, 58, 63, 66
Bones	عظم §§ 2, 44	עצמות §§ 3, 5
Borax	تنكار §§ 6, 22, 43	בורק § 70 טנכר §§ 7, 48, 53
Borax of the Sages	بورق الحكماء § 43	—
Borax salt	—	מלח בורק § 49
Bowl	قدح § 22	קערה § 67
Bread oven	فرن الخبز §§ 3, 12, 14, 32, 34, 46, 48, 49	תנור הלחם §§ 21, 45, 50, 52, 54, 56
Bread Salt	ملح الخبز §§ 2, 6	מלח הלחם §§ 3, 7, 9, 17, 57
Calx	كلس §§ 2, 29, 34, 39	קלס §§ 28, 42, 67 סיד §§ 3, 7, 8, 12, 16, 27, 28, 31, 35, 38, 56, 58, 60
Camphor	كافور § 11	קנפורא/ כנפורא §§ 13, 30
Cinnabar	زنجفر §§ 15, 18, 46, 48	צנברו §§ 18, 24, 30
Clay	—	טיט §§ 24, 29, 45
Clay of the hair	طين الشعر 11§	—

English	Arabic	Hebrew
Clay of the Sages	طين الحكمة/طين المحكم §§ 31, 32, 34, 42, 46	טיט החכמה §§ 12, 13, 17, 25, 26, 35, 44, 48, 50, 56
Cooking pot	قدر/قدرة §§ 3, 5, 7, 8, 32	מזוכך §§ 6, 7, 10, 14, 16, 17, 29, 50, 56, 67, 68
Copper	نحاس §§ 5, 8, 26, 48, 49	נחושת/ נחשת §§ 9, 10, 16, 17, 18, 27, 28, 35, 39, 44, 48, 52, 53, 54, 55, 56, 58, 62, 69, 70 ציפרו § 17
Corporality	تجسد § 5	—
Crocus of iron	—	קרוקו של ברזל §§ 24, 64
Crucible	بوط §§ 27, 34, 41, 42, 43, 44, 48 بودقة §§ 6, 52	אלביט/בוט §§ 37, 40, 48, 49, 51, 53, 56, 64, 70
Crystal	بلور § 5	קריסתלו/ קריסטלו §§ 6, 7
Cypriot vitriol	زاج قبرصي §18	—
Double crucible	بوط بر بوط § 42	—
Dough salt	ملح العجين §§ 4, 7, 13	מלח משי §§ 5, 15
Dragon's blood	دم اخوين § 18	—
Dross	خبث § 42	—
Dryness	يبس §§ 2, 4, 7, 37, 47, 51	יבשות §§ 31, 44
Dust	—	אבק §§ 7, 13, 36, 47
Eagle (sal ammoniac)	عقاب §§ 7, 9, 20, 26, 31, 32, 36, 38, 40, 42, 46, 48, 49, 50	עוקב § 1
Earth	أرض §§ 2, 26, 37	עפר §§ 1, 3, 7, 31, 45
Egyptian eagle (Egyptian sal ammoniac)	عقاب مصري § 34	—
Elixir	إكسير §§ 7, 26, 35, 37, 47	איקציר/אלקציר אדלסיר/אקציר §§ 8, 9, 20, 31, 39, 48, 53, 58, 70

English	Arabic	Hebrew
Elixir/alchemy	كيمياء § 36	—
Exhalation	رائحة § 37	ריח §§ 20, 24, 25, 26, 58, 70
Fat	شحم § 6, 19, 32	—
Ferment	—	חמוץ § 2
	—	שאור §§ 20, 31
Filings	برادة §§ 6, 17, 19, 29, 30, 37, 42, 46, 50	לימטורא/לימטורי לימא §§ 7, 27, 35, 37, 39, 54, 62
Furnace	أتون §§ 28, 29	כור §§ 43, 54
Gemstone	ياقوت § 53	—
Glass	زجاج §§ 4, 20, 42, 43, 51, 52, 53, 54, 55	זכוכית §§ 5, 10, 23, 29, 30, 33, 38, 48, 54, 62, 63, 64, 65, 70
Glazed bowl	—	קערה מזוככת §§ 58, 67
Glazed pot	قدر مزجج / قديرة مزججة §§ 1, 11, 12, 39, 46, 49	קדרה מזוככת §§ 2, 13, 50, 52, 60, 67
Gold	ذهب § 26	זהב §§ 3, 10, 20, 28, 31, 33, 34, 35, 36, 37, 38, 39, 46, 51, 52, 54, 55, 58
Goldsmith's furnace	—	כור הצורפים § 33
Green gemstones	—	יימי ירוקות § 66
Hair	شعر § 18	שער §§ 5, 12, 13
Hotness	حرارة §§ 4, 10, 33, 37, 47	חמימות §§ 28, 29, 31, 50, 58
Humidity	رطوبة §§ 4, 7, 10, 33, 37	לחות §§ 8, 31, 39
Hyacinth	—	יקווינטו § 64
Indian salt	ملح هندي § 2	מלח הנדי § 3

English	Arabic	Hebrew
Iron	حديد §§ 7, 19, 34, 35, 36, 37, 39, 40, 41, 53	ברזל §§ 9, 10, 16, 25, 27, 28, 43, 45, 46, 47, 48, 49, 50, 53, 56, 57, 58, 59, 60, 63, 64, 69
Iron spoon	مغرفة حديد §§ 34, 35, 36, 38, 39, 41	—
Jar	كوز § 48	—
Kitchen salt	ملح الطعام §§ 2, 11, 14	—
Kohl	كحل § 22	—
Ladle	—	תרווד §§ 2, 56, 57, 59, 60, 62, 70
Lapis lazuli	لازورد معدني § 54	אבן אזוורד § 65
Lead	أسرب §§ 14, 26, 28, 33, 36, 37, 38, 39, 40, 41, 42, 43, 44, 46, 48 رصاص §§ 8, 22	אסרוב §§ 25, 58, 60, 61 עופרת/עפרת §§ 17, 23, 25, 26, 33, 36, 39, 39, 46, 47, 48, 49, 53, 58, 59, 62, 69
Lime	جير § 21 نورة § 10	— —
Litharge	مرتك §§ 37, 43	מרתך/מרתק §§ 34, 37, 58 דיתריירו § 49
Magnesia	مغنيسيا §§ 37, 40, 54	מגניסיאה §§ 3, 46, 65
Malachite	دهنج § 55	—
Marcasite	مرقشيثا §§ 22, 40	מרקשיתא §§ 46, 69
Mars (iron)	مريخ §§ 5, 8, 14, 17, 26, 33, 37, 40, 41, 42, 43, 44, 45, 46, 47, 51	—
Medicine	دواء §§ 22, 34, 48, 49	רפואה §§ 9, 20, 30, 51, 52, 56, 64
Mercury/ quicksilver	زيبق §§ 10, 22	כסף חי §§ 12, 17, 20, 28, 67
Milk of the virgin	—	חלב הבתולה §§ 20, 57

English	Arabic	Hebrew
Mineral/ore	معدن/معدنية §§ 4, 10, 33, 37	/מנירים/מינירלי מינירליאה/מעדן §§ 12, 20, 28
Moon (silver)	قمر §§ 2, 8, 14, 25, 26, 28, 32, 33, 34, 37, 40, 43, 46, 47, 48, 50	לבנה §§ 8, 9, 12, 21, 28, 29, 30, 31, 39
Mortar	صلاية §§ 34, 38, 47 مهراس § 49 هاون §§ 14, 48	מכתש §§ 52, 53
Musk	مسك § 24	—
Natron	نطرون § 20	נטרו/נטרון §§ 48, 49, 53, 70
Oil	دهن §§ 4, 7, 10, 14, 42	שמן §§ 23, 37, 44, 47, 48, 50, 70
Oily substance/ ointment	—	אוגיינטו/אוניינטו §§ 2, 5, 9, 20, 28
Operation	تدبير §§ 2, 7, 10, 11, 14, 15, 16, 22, 26, 27, 34, 38, 40, 41, 47	אופירציאוני § 6 קונדימנטו §§ 1, 28, 49, 56, 67, 69
Oven	فرن §§ 17, 41, 42, 48	תנור §§ 1, 4, 14, 16, 21, 41, 47, 48, 59
Red gemstones	—	יימי רוביאי § 65
Red lead	أسرنج § 37	—
Rock salt	—	סל יימא § 3
Ruby	—	רובינו § 64
Saffron	زعفران §§ 18, 46, 53	—
Sal ammoniac	نشادر §§ 6, 7, 20, 23	מלח ארמוניאקו §§ 3, 9 נשדרא §§ 7, 9, 11, 20, 22, 24, 27, 33, 36, 38, 39, 44, 45, 46, 49, 52, 54, 56, 57, 59, 67, 70

English	Arabic	Hebrew
Salt	ملح §§ 2, 4, 5, 6, 7, 9, 11, 12, 25, 28, 29, 30, 31, 34, 36, 38, 39, 48, 50	מלח §§ 3, 5, 7, 8, 9, 10, 11, 13, 14, 16, 20, 21, 22, 27, 28, 32, 33, 35, 36, 38, 41, 42, 43, 44, 48, 53, 54, 55, 56, 57, 59, 60, 61, 62, 66, 68
Salt of Andara	ملح الأندران § 2	—
Salt of ashes	—	מלח האפר § 9
Salt of blood	—	מלח הדם § 3
Salt of calx	—	מלח הסיד §§ 3, 9
Scorpion (sulphur)	عقرب §§ 4, 10, 15, 28, 29, 30, 31, 33, 37, 39, 45, 46, 48, 49	—
Sediment	ثفل §§ 1, 5	שמרים §§ 2, 6, 50, 62
Servant (Mercury/ quicksilver)	عبد §§ 4, 7, 10, 15, 19, 20, 22, 25, 26, 33, 37, 38, 41, 50	—
Sheet	صفحة §§ 13, 31, 32, 41, 43, 44, 45, 46, 48	טס §§ 34, 36, 41, 44, 45, 47, 48, 49, 50, 51, 53, 54, 58, 62
Sign (arsenic)	علم §§ 4, 10, 11, 13, 14, 15, 19, 26, 28, 31, 32, 39, 40, 42, 43, 48	—
Silver	فضة §§ 29, 31, 33, 37	כסף §§ 3, 10, 13, 16, 17, 18, 27, 28, 38, 39, 40, 41, 42, 43, 44, 45, 46, 48, 49, 51, 52, 53, 54, 54, 55, 56, 58, 67, 69, 70
Soap of the sages	صابون الحكماء § 2	ספון החכמים § 3
Soul	نفس §§ 10, 26	נפש §§ 11, 12, 39
Spirit	روح §§ 4, 6, 7, 10, 26, 55	רוח §§ 7, 8, 9, 10, 20, 39
Spirituality	روحانية § 2	רוחות §§ 3, 9, 11, 24, 27, 57, 66
Star (Mercury/ quicksilver)	—	כוכב §§ 3, 8, 9, 18, 20, 21, 22, 23, 24, 25, 26, 27, 31, 33, 36, 38, 39, 44, 45, 47, 48, 49, 54, 55, 56, 58, 59, 63, 67, 70

English	Arabic	Hebrew
Stone	حجر §§ 2, 4, 6, 7, 10, 22, 35	אבן §§ 2, 3, 5, 8, 12, 20, 31, 63, 67
Sulphur	كبريت § 26	גפרית §§ 1, 2, 3, 12, 18, 20, 23, 24, 28, 31, 33, 35, 36, 39, 41, 42, 43, 50, 52, 58, 60 שילפו/שולפו § 50, 52
Sun (gold)	شمس §§ 2, 8, 26, 33, 36, 37, 40, 46, 47, 48, 49	שמש §§ 20, 27, 55
Talc	طلق § 22	טלק § 5
Tin	آنك §§ 21, 26, 39 قلعي §§ 17, 20, 21, 26, 27, 28, 32, 33, 35, 36, 37, 40, 43, 46	אלוק/אלונקי §§ 17, 55, 60, 61 בדיל §§ 3, 10, 11, 16, 17, 20, 27, 28, 33, 39 קלעי §§ 39, 40, 41, 48, 49, 55, 58, 60, 67, 69
Tincture/dye	صبغ §§ 10, 26, 33, 37, 40, 47	צביעה/צבע §§ 1, 2, 24, 39, 46, 54, 55, 58, 63
Turquoise	فيروزج § 54	—
Tutty	توتيا § 18	טוטיאה § 70
Urine	بول §§ 16, 17, 31, 34, 44, 48, 50	שתן §§ 2, 5, 19, 20, 44, 50, 51, 54, 56
Venus (copper)	زهرة §§ 10, 14, 15, 17, 22, 26, 33, 34, 35, 37, 47, 48, 50	—
Verdigris	زنجار §§ 49, 55	זנייר/ זינגייר §§ 52, 66
Vessel	—	כלי §§ 2, 10, 13, 14, 23, 28, 29, 30, 35, 37, 38, 47, 48, 53, 56, 58, 62
Vinegar	خل §§ 6, 10, 13, 19, 34, 36, 38, 39, 46	חומץ §§ 7, 12, 15, 16, 20, 27, 32, 38, 53, 54, 57, 59, 60, 61, 62, 70
Vitriol	زاج §§ 17, 25, 46, 47, 48, 49, 50, 53	זאג §§ 1, 9, 21, 36, 54, 70
Whey	ماء الجبن §§ 6, 16	סיירו §§ 17, 19

English	Arabic	Hebrew
White lead/ ceruse	إسفيداج §§ 37, 43, 44	אסדנג § 58 צירוסי §§ 49, 58, 62

BIBLIOGRAPHY

Manuscripts and Primary Sources

Staatsbibliothek zu Berlin – Preussischer Kulturbesitz, Orientabteilung, MS Orient. Oct. (klein) 514

Staatsbibliothek zu Berlin – Preussischer Kulturbesitz, Orientabteilung, MS Sprenger 1908

Arbuthnott, Catherine J., "Pseudo-Razi De aluminibus et salibus: A Critical Edition and Translation of the Latin Translation with Notes on the Chemical Procedures" (unpublished PhD thesis, University of London, 2002)

Ruska, Julius, *Das Buch der Alaune und Salze: ein Grundwerk der Spätlateinischer Alchemie* (Berlin: Verlag Chemie, 1935)

Steele, Robert, "Practical Chemistry in the 12th Century. Rasis de aluminibus et salibus," *Isis* 12 (1929): 10–49

Books and Articles

Ahlwardt, Wilhelm, *Verzeichniss der arabischen Handschriften der Königlichen Bibliothek zu Berlin*, 10 vols. (Berlin: Asher, 1887–1889)

Anawati, George C., "L'alchimie arabe," in *Histoire des sciences arabes*, ed. Roshdi Rashed, 3 vols. (Paris: Seuil, 1997), III, pp. 111–141

Asin Palacios, Miguel, "Notes to R. Steele's Edition of Rasis *de aluminibus et salibus*," *Isis* 13 (1929–30): 358–9

Bacon, Roger, *Opera hactenus inedita. Opus Minus, Opus Tertium, Compendium Philosophiae* (London: J. S. Brewer, 1859)

Bausi, Alessandro et al., eds., *Comparative Oriental Manuscript Studies. An Introduction* (Hamburg: COMSt, 2015)

Beit-Arié, Malachi, *Hebrew Codicology. Tentative Typology of Technical Practices Employed in Hebrew Dated Medieval Manuscripts* (Jerusalem: The Israel Academy of Sciences and Humanities, 1981; reprint with *addenda* and *corrigenda*, Paris: C.N.R.S., 1977)

Berthelot, Marcellin P. E., *La chimie au Moyen Âge*, 3 vols. (Osnabrück-Amsterdam: Zeller-Philo Press, 1967; reprint of first ed. Paris, 1893)

Bibliotheca Esoterica: catalogue annote et illustre de 6707 ouvrages anciens et modernes, qui traitent des sciences occultes ..., comme aussi des societes secretes (Paris: C. Coulet & A. Faure, 1988; repr. of the ed. Paris: Dorbon-Ainé, 1940)

Blachere, Regis and Jean Sauvaget, *Règles pour éditions et traductions de textes arabes* (Paris: Les Belles Lettres, 1953)

Borellius, Petrus, *Bibliotheca Chimica, seu Catalogus librorum philosophicorum hermeticorum (mit einem Vorwort von Rudolf Schmitz)* (Hildesheim: G. Olms Verlag, 1969; reprint of Heidelberg: Samuel Broun, 1656)

Burnett, Charles, "The Coherence of the Arabic-Latin Translation Program in Toledo in the 12th Century," *Science in Context* 14 (2001): 249–288

Burnett, Charles, "Indian Numerals in the Mediterranean Basin in the Twelfth Century, with Special Reference to the "Eastern Forms," in *From China to Paris: 2000 Years' Transmission of Mathematical Ideas*, eds. Yvonne Dold-Samplonius, Joseph W. Dauben, Menso Folkerts and Benno Van Dalen (Stuttgart: Steiner, 2002), 237–88; reprinted in Burnett, Charles, *Numerals and Arithmetic in the Middle Ages* (London-New York: Routledge, 2010)

Caillet, Albert I. C., *Manuel bibliographique des sciences psychiques ou occultes*, 3 vols. (Paris: Lucien Dorbon, 1912–1913, reprinted in Nieukoop: De Graaf Publishers, 1964)

Carini, Isidoro, *Sulle scienze occulte nel Medio Evo e sopra un codice della famiglia Speciale* (Sala Bolognese: A. Forni, 1983; reprint of Palermo: Perino, 1872)

Cozzando, Leonardo, *Libraria Bresciana* (Sala Bolognese: A. Forni, 1974; reprint of Brescia: Rizzardi, 1694)

Crisciani, Chiara and Jole Agrimi, "Per una ricerca su *experimentum-experimenta:* riflessione epistemologica e tradizione medica," in *Presenza del lessico greco e latino nelle lingue contemporanee*, eds. Pietro Janni and Innocenzo Mazzini (Macerata: Facoltà di Lettere e filosofia dell'Università di Macerata, 1990): 9–49

Crisciani, Chiara, "Fatti, teorie, 'narratio' e i malati a corte. Note su empirismo e medicina nel Tardo Medioevo," *Quaderni di Studi Storici* 108 (2001): 695–717

Crisciani, Chiara, "Meteore, IV: 'Sciant artifices alchimie,'" in *Edizioni, traduzioni e tradizioni filosofiche*, eds. Luca Bianchi, Onorato Grassi, and Cecilia Panti (Canterano RM: Gioacchino Onorati, 2017), 353–367

Dapsens, Marion, "De la *Risālat Maryānus* au *De compositione alchimiae*: Quelques réflexions sur la tradition d'un traité d'alchimie," *Studia graeco-arabica* 6 (2016): 121–40

Dapsens, Marion, "The alchemical works of Khālid ibn Yazīd b. Mu'āwiya (d. c. 85/704), *Asiatische Studien/ Études Asiatiques* 75,2 (2021): 327–427

De Blois, François C., "Al-Ṭughrā'ī," in *Encyclopédie de l'Islam. Second Edition*, eds. Peri Bearman, Thierry Bianquis, Clifford E. Bosworth, Emeri J. van Donzel and Wolfhart P. Heinrichs (Leiden: Brill, 1960–1996), X, pp. 643–644

Della Riviera, Cesare, *Il mondo magico de gli heroi* (Milan: Arché, 1971; expanded reprint of the 2nd ed. Milan: Pietro Martire Locarni, 1605)

Delva, Thijs, "The Abbasid Activist Ḥayyān al-ʿAṭṭār as the Father of Jābir b. Ḥayyān: An Influential Hypothesis Revisited," *Journal of Abbasid Studies* 4 (2017): 35–61

Dietrich, Albert, "Ibn Djuldjul," in *Encyclopédie de l'Islam. Second Edition*, eds. Peri Bearman, Thierry Bianquis, Clifford E. Bosworth, Emeri J. van Donzel and Wolfhart P. Heinrichs, III, 778–779

Dozy, Reinhart P. A., *Supplément aux dictionnaires arabes* (Leiden-Paris: Brill-Maisonneuve, 1927)

Du Fresnoy, Nicolas Lenglet, *Histoire de la philosophie hermétique*, 3 vols. (Paris: Constelier, 1742)

Dupré, Sven, Anna Harris, et al., eds., *Reconstruction, Replication and Re-enactment in the Humanities and Social Sciences* (Amsterdam: Amsterdam University Press, 2020)

Duveen, Dennis I., *Bibliotheca Alchemica et Chemica. An Annotated Catalogue of Printed Books on Alchemy, Chemistry and Cognate Subjects* (Utrecht: HES Publishers, 1986; reprint of London: Weil, 1949 with a contribution by H. P. Kraus, "The Duveen Collection of Alchemy and Chemistry")

Ferguson, John, *Bibliotheca Chemica. A Bibliography of Books on Alchemy, Chemistry and Pharmaceutics*, 2 vols. (Glasgow: J. Macleose, 1906; reprinted in London: Derek Verschoyle Academic and Bibliographical Publications, 1964)

Ferchl, Fritz, *Chemisch-Pharmazeutisches Bio-und Bibliographikon* (Mittenwald: A. Nemayer, 1938)

Ferrario, Gabriele, "Il *Libro degli allumi e dei sali*: *status quaestionis* e prospettive di studio," *Henoch* 26 (2004): 275–296

Ferrario, Gabriele, "Origins and Transmission of the *Liber de aluminibus et salibus*," in Principe, L. (ed.), *Chymists and Chymistry: Studies in the History of Alchemy and Early Modern Chymistry* (Sagamore Beach, MA: Science History Publications, 2007), pp. 137–148

Ferrario, Gabriele, "Understanding the Language of Alchemy: The Medieval Arabic Alchemical Lexicon in Berlin, Staatsbibliothek, MS Sprenger 1908," *Digital Proceedings of the First Annual Lawrence J. Schoenberg Symposium on Manuscript Studies in the Digital Age: Vol. 1: Iss. 1, Article 2*, 2009 (online at: http://repository.upenn.edu/ljsproceedings/vol1/iss1/2)

Ferrario, Gabriele, "An Arabic Dictionary of Technical Alchemical Terms: MS Sprenger 1908 of the Staatsbibliothek zu Berlin (fols. 3r-6r)," *Ambix* 56 (2009): 36–48

Ferrario, Gabriele and Maud Kozodoy, "Sciences and Medicine", in *The Cambridge History of Judaism*, Vol. 5: 'The Middle Ages', ed. Philip Ackerman-Lieberman (Cambridge: Cambridge University Press, 2021), pp. 825–863

Fors, Hjalmar, Lawrence M. Principe and H. Otto Sibum, eds., "From the Library to the Laboratory and Back Again: Experiments as a Tool for Historians of Science," *Ambix* 63.2 (2017)

Forster, Regula, "Jābir ibn Ḥayyān" in *Encyclopaedia of Islam* 3, eds. Kate Fleet, Gudrun Krämer, Denis Matringe, John Nawas and Everett Rowson (Leiden-Boston: Brill, 2017), 91–97

Frati, Lodovico, *Indice dei codici latini conservati nella R. Biblioteca Universitaria di Bologna* (Florence: Successori Seeber, 1909; reprint of the article first published in *Studi italiani di filologia classica* 16 [1908]: 103–482)

Freudenthal, Gad, ed., *Science in Medieval Jewish Cultures* (Cambridge: Cambridge University Press, 2012)

Gacek, Adam, *The Arabic Manuscript Tradition. A Glossary of Technical Terms and Bibliography–Supplement* (Leiden-Boston: Brill, 2001)

Gacek, Adam, *The Arabic Manuscript Tradition. A Glossary of Technical Terms and Bibliography* (Leiden-Boston-Köln: Brill, 2001)

Gacek, Adam, *Arabic Manuscripts. A Vademecum for Readers* (Leiden-Boston: Brill, 2009)

Gherli, Fulvio, *Proteo Metalico o sia delle trasformazioni superficiali de'Metalli e delle differenti preparazioni de'medesimi molto proprie per debellare i mali più atroci, che il Corpo Umano affliggono, e per iscoprire gl'inganni de'falsi Chimici* (Venezia: G. Corona, 1721)

Gmelin, Johann Friedrich, *Geschichte der Chimie seit dem Wiederaufleben der Wissenschaften bis an das Ende des 18. Jahrhunderts*, 3 vols. (Hildesheim: G. Olms Verlag, 1965; reprint of Göttingen: J.G. Rosenbusch, 1797–1799)

Gutas, Dimitri, *Greek Thought, Arabic Culture: the Graeco-Arabic Translation Movement in Baghdad and Early 'Abbāsid Society* (Oxford: Routledge, 1998)

Hall, Martin, *John of Garland's* De triumphis Ecclesie. *A new critical edition with introduction and translation* (Turnhout: Brepols, 2020)

Hallum, Benjamin C., "The *Tome of Images*: an Arabic Compilation of Texts by Zosimos of Panopolis and a Source of the *Turba Philosophorum*," *Ambix* 56 (2009): 76–88

Haskins, Charles H., *The Renaissance of the Twelfth Century* (Cambridge: Harvard University Press, 1927)

Heym, Gerard, "Al-Razi and Alchemy," *Ambix*, 1 (1938): 184–191

Hoefer, Ferdinand, *Histoire de la chimie depuis les temps les plus reculées jusqu'a notre époque*, 2 vols (Paris: Hachette, 1842–43)

Holmyard, Eric J., *The Arabic Works of Jābir ibn Ḥayyān* (Paris: Geuthner, 1928)

Hugonnard-Roche, Henri, "Les traductions du grec au syriaque et du syriaque à l'arabe," in *Rencontres de cultures dans la philosophie médiévale. Traductions et traducteurs de l'antiquité tardive au XIV*[e] *siècle*, eds. Jacqueline Hamesse and Marta Fattori (Louvain-la-Neuve-Cassino: Institut d'études médiévales, 1990), pp. 131–148

Ibn, Ǧulǧul, *Les generations du médicins et des sages*, ed. Fu'ād Sayyid (Frankfurt am Main: Institute for the History of Arabic-Islamic Science, 1996; reprint of the edition Al-Qāhira, 1955)

Irani, Rida A. K., "Arabic Numeral Forms," *Centaurus* 4 (1955): 1–12

Ivry, Alfred L., "Philosophical Translations From the Arabic in Hebrew During the Middle Ages," in *Rencontres de cultures dans la philosophie médiévale. Traductions et traducteurs de l'antiquité tardive au XIV*[e] *siècle*, eds. Jacqueline Hamesse and Marta Fattori (Louvain-La-Neuve/Cassino: Institut d'études médiévales, 1990), 167–186

Al-Jiwasī, Mustafa, *Mawsu'a 'ulamā' al-'arab wa-al-muslimīn wa-a'lāmihim* ('Amman: Dār Asmā' li-nashr wa-al-tawzī,' s.d.)

Kahn, Didier, "Le fonds Caprara de manuscrits alchimique de la Bibliothèque universitaire de Bologne," *Scriptorium* 48 (1994): 62–110

Käs, Fabian, *Die Mineralien in der arabischen Pharmakognosie. Eine Konkordanz zu mineralischen Materia medica der klassischen arabischen Heilmittelkunde nebst überlieferungsgeschichtlichen Studien* (Wiesbaden: Akademie der Wissenschaften und Literatur, 2010)

Kopp, Hermann, *Ansichten über die Aufgabe der Chemie und über die Grundbestandtheile der Körper bei den bedeutenderen Chemikern* (Braunschweig: Vieweg, 1875)

Kopp, Hermann, *Die alchemie von letzten Viertel des 18. Jahrhunderts: die Alchemie in älterer und neuerer Zeit*, 3 vols. (Hildesheim-New York: G. Olms Verlag, 1971; reprint of Heidelberg, 1886)

Kraus, Paul, *Jābir ibn Ḥayyān. Textes choisis* (Paris/Al-Qāhira: Maisonneuve/El-Khandgi, 1935)

Kraus, Paul, "Julius Ruska. Festschrift zum 70. Geburstag," *Osiris* 5 (1938): 4–40

Kraus, Paul, *Jābir ibn Ḥayyān. Contribution à l'histoire des idées scientifiques dans l'Islam*, 2 vols. (Al-Qāhira: Imprimerie de l'Institut Français d'Archéologie Orientale,1942–43; repr. Paris: Le Belles Lettres, 1988)

Langermann, Tzvi Y., *The Jews and the Sciences in the Middle Ages* (Aldershot: Ashgate Publishers, 1999)

Leclerc, Lucien, *Traité des Simples par Ibn el-Beïthar. Notices et extraits des manuscrits de la Bibliothèque Nationale* (Paris: Imprimerie Nationale, 1883)

Lemay, Richard, "Arabic Numerals," in *Dictionary of the Middle Ages*, ed. Joseph Strayer (New York: Charles Scribner's Sons, 1982), 382–98

Lévi-Provençal, Évariste and Charles Pellat, "Al-Maḳḳarī," in *Encyclopédie de l'Islam. Second Edition*, eds. Peri Bearman, Thierry Bianquis, Clifford E. Bosworth, Emeri J. van Donzel and Wolfhart P. Heinrichs (Leiden: Brill 1960–96), VI, 170–2

Lewicki, Tadeusz, "Al-Ḳazwīnī," in *Encyclopédie de l'Islam. Second Edition*, eds. Peri Bearman, Thierry Bianquis, Clifford E. Bosworth, Emeri J. van Donzel and Wolfhart P. Heinrichs (Leiden: Brill, 1960–96), IV, 898–900

Lory, Pierre, *Jābir ibn Ḥayyān. Dix traités d'alchimie. Le dix premiers traités du 'Livre de soixantedix'* (Paris: Sindbad, 1983)

Lory, Pierre, *Tadbīr al-iksīr al-a'ẓam – L'elaboration de l'elixir supreme* (Damas: Institut Français de Damas, 1998)

Maas, Paul, *Textkritik*, 2nd ed. (Leipzig: Teubner Verlagsgesellschaft, 1950)

Al-Maqqarī, *Analectes sur l'histoire et la littérature des Arabes d'Espagne*, eds. Reinhart P.A. Dozy, Gustave Dugat, Ludolf Krehl and William Wright (Leiden: Brill, 1855–61)

Marlow Taylor, Gail, *The Alchemy of Al-Razi. A Translation of the Book of Secrets* (Self-published and printed, 2014)

Martelli, Matteo, *Pseudo-Democrito, Scritti alchemici, con il commentario di Sinesio. Edizione critica del testo greco, traduzione e commento* (Milan-Paris: Arché-SÉHA, 2011)

Martelli, Matteo, *The* Four Books *of Pseudo-Democritus* ("Sources of Alchemy and Chemistry" 1) (Leeds: Maney Publishing, 2013)

Martelli, Matteo, Sébastien Moureau and Jennifer Rampling, "Theory and Concepts: The Shared Heritage of Byzantine, Arabo-Muslim, and Latin Alchemy," in *A Cultural History of Chemistry in the Middle Ages*, eds. Charles Burnett and Sébastien Moureau (London: Bloomsbury, 2022), 18–23.

Moureau, Sébastien, "Les sources alchimiques de Vincent de Beauvais," *Spicae. Cahiers de l'Atelier Vincent de Beauvais* (New Series) 2 (2012): 5–118

Moureau, Sébastien, *Le* De Anima *alchimique du pseudo-Avicenna*, 2 vols. (Firenze: Sismel-Edizioni del Galluzzo, 2016)

Moureau, Sébastien and Nicolas Thomas, "L'aludel: savoir et savoir-faire transmis du monde arabe à l'Occident médiéval?," in *Héritages arabo-islamiques dans l'Europe méditerranéenne*, ed. Catherine Richarté and Roland-Pierre Gayraud Jean-Michel (Paris, La Découverte, 2015): 239–252.

Moureau, Sébastien and Nicolas Thomas, "Practice and Experiment: Alchemical Operations in the Middle Ages," in *A Cultural History of Chemistry: The Middle Ages*, eds. Charles Burnett and Sébastien Moureau (London: Bloomsbury, 2022), 35–47

Al-Nadīm, *Kitāb al-Fihrist*, ed. Gustav L. Flügel, 2 vols. (Leipzig: Vogel, 1871–1872)

Al-Nadīm, *The Fihrist of Al-Nadīm. A Tenth Century Survey of Muslim Culture*, ed. Bayard Dodge, 2 vols. (New York/London: Columbia University Press, 1970)

Nazari, Giovanni Battista, *Della tramutazione metallica sogni tre* (followed by the *Canzone Alchemica* by Rigino Danielli da Capodistria) (Milan: Arché, 1967; reprint of the 2nd ed. Brescia: Marchetti, 1599)

Newman, William R. and Lawrence M. Principe, "Alchemy vs. Chemistry: The Etymological Origins of a Historiographic Mistake," *Early Science and Medicine* 3 (1998): 32–65

Nomanul Haq, Syed, *Names, Natures and Things. The Alchemist Jābir ibn Ḥayyān and his* Kitāb al-Aḥjār (New York: Springer, 1994)

Norris, John, "The Mineral Exhalation Theory of Metallogenesis in Pre-Modern Mineral Science," *Ambix* 53 (2006): 43–65

Obrist, Barbara, *Les débuts de l'imagerie alchimique (XIVe-XVe siècles)* (Paris: Éditions le Sycomore, 1982)

Partington, James R., "The Chemistry of Al-Razi," *Ambix*, 1 (1938), pp. 192–196

Patai, Robert, *The Jewish Alchemists. A History and Source Book* (Princeton: Princeton University Press, 1994)

Pereira, Michela, *Alchimia. I testi della tradizione occidentale* (Milan: Mondadori, 2006)

Peroni, Vincenzo, *Biblioteca Bresciana*, 3 vols. (Sala Bolognese: A. Forni, 1968; reprint of Brescia, 1816–1823)

Principe, Lawrence M., *The Secrets of Alchemy* (Chicago: University of Chicago Press, 2013)

Al-Qazwīnī, *Géographie/Kosmographie: Aṯār al-Bilād*, ed. Ferdinand Wüstenfeld (Göttingen: Verlag der Dieterischen Buchhandlung, 1848)

Quattrami, Evangelista, *La vera dichiarazione di tutte le metafore, similitudini e enimmi de gl'antichi Filosofi Alchimisti, tanto Caldei e Arabi, come Greci e Latini, usati da loro nella descrittione, e compositione dell'Oro potabili, Elissire della vita, Quinta essenza e Lapis filosofico ... si mostra l'errore e ignoranza (per non dir l'inganno) di tutti gli Alchimisti moderni* (Roma: V. Accolti, 1587)

Rampling, Jennifer M., "Transmuting Sericon: Alchemy as 'Practical Exegesis' in Early Modern England," *Osiris* 29 (2014): 19–34

Rampling, Jennifer M., "How to Sublime Mercury: Reading Like a Philosopher in Medieval Europe." *The Recipes Project* (24 May 2018), https://recipes.hypotheses.org/10663

Rampling, Jennifer M., *The Experimental Fire: Inventing English Alchemy, 1400–1700* (Chicago: University of Chicago Press, 2020)

Rampling, Jennifer M., "Art and Representation: The Alchemical Image in the Islamic and Christian Middle Ages," in *A Cultural History of Chemistry: The Middle Ages*, eds. Charles Burnett and Sébastien Moureau (London: Bloomsbury, 2022), 149–178

Ranking, George S. A., "The Life and Works of Rhazes," in *Proceedings of the XVIIth International Congress of Medicine* (1913), *Section of the History of Medicine* (Oxford, 1914), 246–268; reprinted in Fuat Sezgin, *Muhammad ibn Zakariya al-Razi (d. 313/925): Texts and Studies* (Frankfurt am Main: Institute for the History of Arabic-Islamic Science at Johann Wolfgang Goethe University, 1996)

Roberts, Julian and Andrew G. Watson, *John Dee's Library Catalogue* (London: Bibliographical Society, 1990)

Rothschild, Jean-Pierre, "Du Latin à l'Hébreu: quelques problèmes posés par des traductions médiévales," in *Rashi 1040–1990. Hommage à Ephraïm E. Urbach*, ed. Gabrielle Sed-Rajna (Paris: Les Èditions du Cerf, 1993)

Roth-Scholtz, Friedrich, *Deutsches Theatrum Chemicum*, 3 vols. (Hildesheim-New York: G. Olms Verlag, 1976; reprint of the ed. 1732)

Ruska, Julius, "Kazwini-Studien," *Der Islam* 4 (1913): 14–66; 236–62

Ruska, Julius, "Al-Razi (Rhazes) als Chemiker," *Zeitschrift für Angewandte Chemie*, 35 (1922): 719–724

Ruska, Julius, "Al-Razi als Bahnbrecher einer neuer Chemie," *Deutsche Literatur-Zeitung*, 44 (1923): 117–124

Ruska, Julius, *Arabische Alchemisten I. Chālid ibn Jazid ibn Mu'āwija* (Heildelberg: Akten der von-Portheim-Stiftung 6, 1924)

Ruska, Julius, "Übersetzung und Bearbeitung von al-Razi's *Buch Geheimnis der Geheimnisse*," in *Quellen und Studien zur Geschichte der Naturwissenschaften* IV (Berlin: Springer, 1935)

Ruska, Julius, "Die Alchemie ar-Razi's," *Der Islam* 22 (1935): 281–319

Ruska, Julius, "Pseudoepigraphe Rasis-Schriften," *Osiris* 7 (1939): 31–94

Schmieder-Jappe, Thomas, *Die Sammlung der orientalischen Handschriften der Staatsbibliothek zu Berlin* (Berlin: Logos Verlag, 2004)

Sezgin, Fuat, *Geschichte des Arabischen Schriftums* (Leiden: Brill, 1971)

Sharbil, Maurice, *Mawsu'a al-Muktashifīn wa-al-Mukhtari'īn* (Bayrūt: Dār Al-Kutub Al-'Ilmiyya, 1991)

Siggel, Alfred, *Katalog der arabischen alchemistischen Handschriften Deutschlands: v. I: Handschriften der öffentlichen wissenschaftlichen Bibliothek – früher Staatsbibliothek Berlin*, (Berlin: Akademie Verlag 1949)

Siggel, Alfred, *Decknamen in der arabischen alchemistischen Literatur* (Berlin: Akademie-Verlag, 1951)

Singer, Dorothea W., *Catalogue of Latin and Vernacular Alchemical Manuscripts in Great Britain and Ireland*, 3 vols. (Brussels: Lamertin, 1928)

Stavenhagen, Lee, *A Testament of Alchemy: Being the Revelations of Morienus to Khālid ibn Yazid* (Hanover, NH; Brandeis University Press, 1974)

Steinschneider, Moritz, *Die Handschriften-Verzeichnisse der Königlichen Bibliothek zu Berlin* (Hildesheim/New York: G. Olms Verlag, 1980; reprint of Berlin: Buchdruckerei der König. Akademie der Wissenschaften, 1878)

Steinschneider, Moritz, *Die Europäischen Übersetzungen aus dem Arabischer bis mitte des 17 Jahrhunderts* (Graz: Akademische Druck- u. Verlagsanstalt 1956; repr. of the ed. Wien: Sitzungsberichte Akademie der Wissenschaften in Wien, phil.-hist. Klasse, 101, 1, 1905)

Thomas, Nicolas and Caroline Claude, "Les Vases à fond percé: pratique de la distillation *per descensum* au bas Moyen Âge en Île-de-France," *Revue Archéologique d'Île-de-France* 4 (2011): 267–288

Thomas, Nicolas, "De la recette à la pratique, l'exemple du lutum sapientiae des alchimistes," in *Craft Treatises and Handbooks: the Dissemination of Technical Knowledge in the Middle Ages*, ed. Ricardo Cordoba de la Llave (Turnhout: Brepols, 2013), 249–70

Thomas, Nicolas and Sébastien Moureau, "Laboratories and Technology: Alchemical Equipment in the Middle Ages," in *A Cultural History of Chemistry: The Middle Ages*, eds. Charles Burnett and Sébastien Moureau (London: Bloomsbury, 2022) 49–70

Thorndike, Lynn, *A History of Magic and Experimental Science*, 8 vols. (New York/London: Columbia University Press, 1923–1958)

Ullmann, Manfred, *Die Natur- und Geheimwissenschaften im Islam* (Leiden: Brill, 1972)

Ullmann, Manfred, "Ḫālid ibn Yazīd und Alchemie: Eine Legende," *Der Islam* 55 (1978): 181–218, reprinted in *Aufsätze zur arabischen Rezeption der griechischen Medizin und Naturwissenschaften*, ed. Rüdiger Arnzen (Boston/Berlin: De Gruyter, 2016), 334–70

Unione Accademica Nazionale, *Catalogo dei manoscritti filosofici delle biblioteche italiane*, 11 vols., (Firenze: Olschki, 1993)

Weil, Gustav and Georges S. Colin, "Abdjad," in *Encyclopédie de l'Islam. Second Edition*, eds. Peri Bearman, Thierry Bianquis, Clifford E. Bosworth, Emeri J. van Donzel and Wolfhart P. Heinrichs (Leiden: Brill 1960–1996), I, 97–8

Wilson, William J., "Catalogue of Latin and Vernacular Alchemical Manuscripts in the United States and Canada," *Osiris* 6 (1939): 419–461

Winderlich, Rudolf, "Ruska's researches on the alchemy of Al-Razi," *Journal of chemical education*, 13 (1936): 313-315

Wright, Thomas, *De Triumphis Ecclesiae Johannis de Garlandia* (London: Nichols, 1856)

INDEX OF NAMES OF PERSONS, PLACES, AND TITLES OF WORKS